FRESHWATER AQUACULTURE

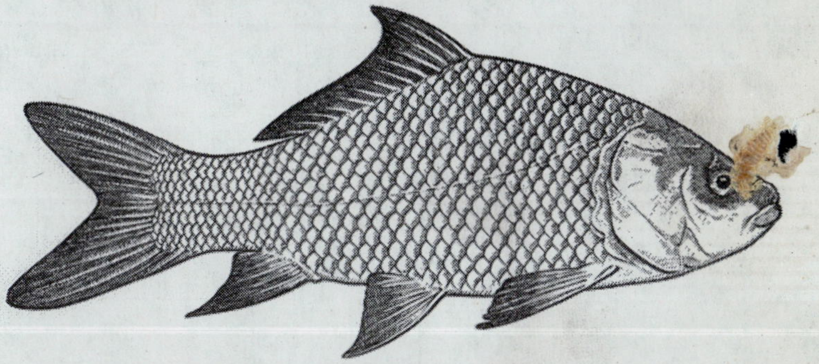

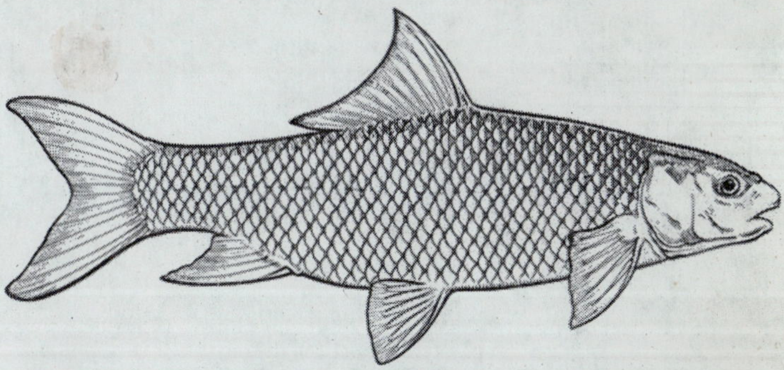

FRESHWATER AQUACULTURE

Dr. S.H. Ahmad
Dr. A.K. Singh

2011
DAYA PUBLISHING HOUSE®
Delhi - 110 002

ISBN (PB) 81-7035-709-8
ISBN (PB) 978-81-7035-709-4

Published by : **Daya Publishing House®**
4760-61/23, Ansari Road, Darya Ganj,
New Delhi - 110 002
Phone: 23245578, 23244987
Fax: (011) 23260116
e-mail : dayabooks@vsnl.com
website : www.dayabooks.com

Laser Typesetting : **Classic Computer Services**
Delhi - 110 035

Printed at : **Chawla Offset Printers**
Delhi - 110 052

PRINTED IN INDIA

URMILA SINGH
Governor
Himachal Pradesh

सत्यमेव जयते

RAJ BHAVAN,
SHIMLA - 171 002

Foreword

In recent days, Freshwater Aquaculture has made tremendous advances consequently producing large variety of fish as source of food. This, in turn has revolutionized the knowledge of freshwater aquaculture science besides meeting the increasing global demand of sea-food. On the other hand it also provides large scale employment to the poor fishermen.

Dr. Syed Hasib Ahmad, former Associate Dean-cum-Principal, College of Fisheries, Dholi (Rajendra Agricultural University), Pusa, Samastipur (Bihar) and former Advisor Aquaculture and Fisheries, Department of Institutional Finance and Programme Implementation, Government of Bihar, Patna is an eminent scholar with varied experience of teaching research extension in aquaculture and fisheries. Dr Arun Kumar Singh. Lecturer in Zoology, B.D. College (Magadh University), Patna, another author has also contributed a good number of research papers and articles in the field of aquaculture and fisheries.

I hope that the book entitled *"Freshwater Aquaculture"* jointly authored by Dr. Ahmad and Dr Singh will be well received by the teachers, students, extension workers as well as by those who are engaged in commercial exploration of aquaculture and fisheries.

Urmila Singh

Urmila Singh

Preface

In India, however, significant achievement have been made in freshwater carp culture at all levels, from technology development to technology demonstration and progressive development. Since 1979, the growth of the freshwater aquaculture industry has been tremendous, but the increase in production of crustaceans shrimps prawns and crabs the expansion of brackishwater aquaculture have been equally much more impressive, our purpose is to provide a practical discussion of water quality principles and management for use by aquaculturists, farmers, students, teachers, researchers and the other persons involved in related activities to boost aquatic production.

The term water quality in its broadest perspectives includes all physical, chemical and biological characteristics of water and its direct or indirect impact on aquaculture. However, now-a-days aquaculturists are concerned with those aspects of water quality which regulate the suitability of water for holding or rearing aquatic organisms. Because failure to maintain proper water quality in an environment more particularly so in a pond may cause poor growth or death of aquatic animals, thus aquaculturists should have a basic and thorough knowledge and understanding of water quality. Unfortunately, many do not have the physical, chemical, and also the mathematical background required to appreciate many of the theoretical works on water quality and its management planning for enhancing the production and proper maintaining sustainable aquaculture without the deterioration of the aquatic environment.

It is always advisable to culture the innumerable types of freshwater organisms of economic values in the various types of freshwater bodies in a sustainable manner without disturbing its ecology as well as without deteriorating the water quality parameters to boost aquaculture production to the maximum sustainable yield by the planners, aquaculturists, fisheries scientists, extension workers, farmers, and scientists.

We express our thanks and gratitude to M/S Daya Publishing House, Delhi, who took interest in this venture and book could see the light of the day.

Dr. S.H. Ahamd

Dr. A.K. Singh

Contents

Chapter 1
Introduction

Fish culture in ponds, has been practiced in India since as early as 350 B.C. The bio-industry apparently became more popular in later years, as is seen from old records which show that even temple tanks were regularly stocked with fish seed and their fishing rights leased out on annual basis.

Fish culture is very popular in the States of Bihar, West Bengal, Orissa and Uttar Pradesh. In other states, too, there has been marked awakening of interest in inland fisheries.

At present, about 2,46,867 acres (1 hectare = 2.471 acres) of inland waters are under fish culture in India. In India as a whole irrigation canals exceed 70,000 miles (1 mile = 5,280 ft) in length of wells 13.5 million in number. Ponds and tanks are much more numerous in West Bengal, Bihar and in Orissa.

Cultivable waters are of all types, perennial or seasonal, and of every description-ponds and tanks, natural lakes and artificial reservoirs, backwaters, irrigation canals and wells, paddy fields, jhils, bheels, borrow-pits, quarry pools, swamps, and even short-season accumulations of water in shallow low lying areas.

There is little or no natural fishery in these waters. A systematic cultivation of quick-growing species of fish in the extensive, widely distributed inland waters will be one good way of ensuring an adequate supply, in the fresh conditions, of a cheap, wholesome protective food containing high quality animal protein which is badly lacking in the diet of our people.

The systematic cultivation of quick-growing fish all cultivable inland waters is thus the ultimate aim of inland fisheries development in India more particularly so in landlocked states.

Indian major carps (Rohu, Catla and Mrigal) form the most important group of fishes used for fish culture. Other species, including the predatory murrels (*Channa* spp.), and selected salt-water species such as Pearl Spot and Milk Fish are also cultivated on a limited scale in certain areas in India.

The larger Indian carps live in rivers and spawn during the monsoon months of June to August, when millions of eggs, hatchlings and fry are collected for stocking in ponds and tanks, particularly in West Bengal, Bihar, Orissa and Assam. In West Bengal, there is a flourishing seasonal fish fry trade. Fishermen collect the fry in large numbers in special fry-catching nets and transport them in earthen hundies by rail to Kolkata for sale. Pond owners purchase their fish seed requirements from the market and release them in their ponds.

In several states in India, Indian major carps (IMC) are not available, and quick-growing local species suitable for cultivation are few in number. These states, therefore import the seed for the Indian major carps mainly from West Bengal, Orissa and Bihar every year. Apart from the open market, fish supplies are also obtained from the Fish Seed Syndicate, organized under the auspices and technical guidance of the Central Inland Fisheries Research Station, Kolkata. In Chennai, Andhra Pradesh and Punjab, Carp fingerlings are collected from rivers, canals and paddy fields for stocking fish ponds and tanks.

Collecting the fry and rearing them in nursery ponds is a paying profession, but yet to be developed in these states. Most fish farmers employ empirical cultural practices. They feel that they have done their duty when they have released a cupful of fry in the pond, and wait for the proper season to reap the harvest.

Due to faulty methods of collecting and handling, the tiny fry suffer serious mortality from the stage of catching them in the rivers to that of rearing them in ponds. Ponds are not properly cleared of predatory animals before introducing the small fish. Adequate steps are not taken, due mainly to ignorance, to ensure sufficient food for the young fish. Ponds are indiscriminately stocked. Even the expert professional fish farmer, therefore, are never sure of a successful crop of fry of fingerlings in his ponds. Total failures of the fry to survive in such nursery ponds are not uncommon. Only a very small fraction of the millions of fry collected grow up to a marketable size.

Cultivable Waters

India has a large variety of inland waters, ranging from the ice cold Himalayan streams and lakes to the tropical warm water ponds. In all these waters there is life, and in the majority of them a natural fish fauna also. Fish production in these waters, however, is comparatively small.

The extent of these waters, though not assessed properly yet, is indeed enormous, and with the number of large multipurpose river projects nearing completion, the area of the inland cultivable waters will increase by several million hectares.

There are vast swamps and other low-lying accumulations of water which only act as breeding grounds for mosquitoes, thus endangering public health. Converting them into productive fish farms will not only check water-borne disease, but also increase fish production in the country. Continued neglect of even cultivated waters often results in an overgrowth of aquatic weeds in them. This changes the quality of waters and results in an overgrowth of aquatic weeds in them. With the adoption of scientific methods of culture any piece of water can be made to yield a better crop of fish.

Indian waters vary widely in the different parts of the country owing to difference in the soil and rainfall and other biophysical factors. Hence, they vary in their productivity. The water on acid soils, for example, is generally less productive than that on alkaline soils. Acid soils are mostly found in certain parts of Assam and in some of the upland hilly areas. A recent survey shows that hardly 7 per cent of the waters in Uttar Pradesh are acid.

The majority of our inland waters are alkaline, extremely alkaline conditions which are not too good for optimum fish production are also met with in some waters of the dry belt region. A total alkalinity of over 500 parts per million (ppm.) has been recorded in certain waters in Andhra Pradesh.

Fish production in ponds depends of biological factors also. A pond with abundant submerged weeds may harbour a variety of life like, insects, snails, worms and small fishes, but the production of edible fish in such a ponds, almost every time, is poor.

At present, very little information is available on the natural production of fish in our waters. The inland waters may be classified into the following groups.

Table 1.1

Pond Type	*Nature of Food Organisms*	*Colour of Water*	*pH*	*Nature of Pond Bottom*	*Average Yield of Fish (kg//ha/year)*
I.	Permanent bloom of Blue green algae like *Microcystis, Anabaena,* or *Oscillatoria*	Greenish	8.0	Black rotting organic matter	5,000
II.	Abundant macro-vegetation, *Hydrilla, Vallisneria, Potomogeton,* water lilies, etc.	Colourless	7.5 to 8.5	–do–	2,500
III.	Abundance of zooplankton and a few plant organisms	Brownish	7.0	–do–	1,100
IV.	Clear throughout with an occasional, bloom	Clear	6.5 to 7.5	Brownish	660
V.	Few phytoplankton and zooplankton	Clear	6.5 to 7.5		450

The above classification, though very general, gives a rough idea of the yield that can be expected even from the best type of water. In West Bengal and Bihar the production of major carps (Catla, Rohu and Mrigal) is reported to vary from 235 to 617 kg per ha, with supplementary (artificial) feeding it is about 864 to 1,361 kg per ha. But the properly selecting fast-growing fish, stocking them in optimum densities and combinations of systematic supplementary feeding, this yields can definitely be increased.

Chapter 2

The Typical Fish Pond

In a typical fish pond, the water is essentially stagnant (still). Water constitutes the immediate environment of all aquatic organisms. The dissolved substances and nutrients contained in the water undergo changes which affect the conditions of existence of the aquatic organisms.

For the successful rearing of fish, the water should not only contain the necessary energy and nutrients in the optimum concentrations, but also food organisms of the fish. Certain basic elements like carbon, hydrogen, sulphur, nitrogen, phosphorus, potassium, calcium, magnesium and iron should be present in certain proportions. Sodium, chlorine, fluorine, silicon, manganese, iodine and arsenic are also essential for fish growth. These substances are obtained from the air, land the soil which are in continuous contact with water. Thus, the water conditions will largely depend on the nature of the soil on which the fish pond is constructed.

In each pond, various groups of plant and animal organisms, inhabiting particular niches are present. The life activities of these are complicated. There are two important phases of activity in the life history of any organisms. The constructive and the destructive-taking place almost simultaneously. For the productive phase, the ultimate basic substances in the pond are the inorganic nutritive materials dissolved in water and certain energies (light and heat) and gases (oxygen and carbon dioxide) form the atmosphere.

The green coloured microscopic plants that float in water (phytoplankton), the submerged larger plants and certain bacteria synthesise these inorganic nutrients into complex living matter in the presence of sunlight and dissolved nutrients found in the mud. The microscopic free-floating plants in water are eaten by minute animals, the zooplankters. Both these groups of organisms are directly consumed by several species of fishes. The submerged plants, both in the fresh and in the decaying conditions are not only eaten directly by certain species of fish, but they also harbour a host of herbivorous and creeping animals like the snails and insect larvae which, in turn, are preyed upon by predatory animals.

A large number of organisms, collectively termed as the bottom fauna (benthos), feed on the bottom debris. These form the food of the fishes and other carnivorous animals in water. The latter which feed also on the free swimming animalcules (zooplankton), are in their turn fed upon by predatory fishes which grow and multiply. This building up phase in the pond culminates in the production of fish.

At each stage of this constructive phase, there is also an active destructive cycle, the different organisms being either consumed as food by the larger organisms, or dying a natural death. After death, they settle down at the bottom as organic debris and are gradually decomposed by bacterial action into the component inorganic substances or eaten by bottom animals and bottom-feeding fishes.

The cycle of life in pond is thus a continuous process of building up of substances into more complex ones and a process of breaking up of these substances into simpler ones through bacterial action. In a cultivated ponds, we help to stimulate the different phases of the constructive cycle by cultural methods and periodically harvest the ultimate final product, the fish.

The vital life-processes in the pond are at the optimum, when the physical and chemical conditions of water are normal. The important physical and chemical conditions of water which have a great effect on fish production are given below.

Depth

When the water is shallow, light will penetrate even to the bottom, warm up the water and facilitate increased production. When the water is too deep, the bottom layers will be cold and less productive, and will gradually accumulate poisonous gases which reduce normal production. A water less than three feet in depth, however, is likely to get heated up too much on hot, sunny days.

Transparency/Visibility

The turbidity of water may be due to the abundance of plankton or to the suspension of fine particles of silt. When the water is too turbid, light penetration is interfered with, resulting in reduction in biological production.

Temperature

This is one of the most important factors. The temperature varies at different times of the day and also during different seasons of the year, from place to place. In tropical waters, biological production and growth are much more rapid than in temperate waters. Warm water fishes are, therefore, ill-adapted to life in cold waters, a fact to be remembered when fishes are selected for stocking. In certain parts of India where the summer is very hot and the winter correspondingly cold, the amplitude of water temperature is sometimes as much as 20° C or more. The Indian major carps (IMC) usually tolerate this wide range of temperature, but during winter their growth is very much retarded.

Dissolved Oxygen (DO)

Aquatic animals get their supply of oxygen for breathing from the air dissolved in water. The specialized respiratory structures called gills in fishes are capable of absorbing oxygen from the water and giving out waste gases (carbon dioxide) from the blood into the water. As the water surface is always in contact with the air, oxygen gets dissolved in water almost continuously. The photosynthetic activity of water-plants also enriches the water with oxygen.

An ordinary pond will generally contain sufficient dissolved oxygen to support its fish population. When weeds and phytoplankton are abundant, the dissolved oxygen (DO) in water undergoes wide fluctuations, sometimes even reaching dangerous limits during the night hours. Oxygen is bound to be deficient when the water is highly contaminated with organic matter.

Different fishes need different quantities of dissolved oxygen. The common trout requires much more oxygen than the Indian Major Carp (IMC). While the oxygen consumption of fish depends on different environmental factors, a concentration of 3 to 5 parts per million (ppm.) of water is sufficient for most of our indigenous fishes. Under certain conditions, our Indian Major Carps (IMC) have been found to tolerate even a much lower concentration of dissolved oxygen.

When the dissolved oxygen in water is not sufficient for respiratory purposes, the fish comes up in distress to the surface to gulp the atmospheric air, and if not transferred to fresh, oxygenated water, gets easily asphyxiated.

Carbon Dioxide

Though essential for the photosynthetic activity of the plants, carbon dioxide is a waste product for fish. The gas, which is dissolved in water from the air, is often abundant in the subsoil spring water. Considerable quantities of this gas are formed when there is excessive organic decomposition like the rotting of plants. In all natural waters, some carbon dioxide is usually present, but even up to 15 to 20 parts per million (ppm) it has no adverse effect on fish life. It becomes dangerous only when this concentration is exceeded.

Acidity, Alkalinity and Neutrality of Water

Natural water, like any other liquid, may be acid, alkaline or neutral. The reaction is expressed in terms of water is called "pH" value. When the reaction is neutral, it is generally expressed as pH–7. When the pH is above 7.0, the water is alkaline in reaction and when the pH is below 7.0, the water is acidic in reaction.

The majority of natural waters are alkaline. Acid waters pH below 6.0 are much less productive than alkaline waters, and the fish in such waters often get diseased and gradually die.

Alkalinity of the water is mainly due to calcium salts in the form of bicarbonates and carbonates. In highly productive waters, the bicarbonate content will ordinarily be high, over a 100 parts per million (ppm) of water.

Dissolved Nutrients

Phosphates and nitrates are the important nutrients necessary for the production of fish and fish food in water. In all natural waters, three nutrients are found in varying quantities, dissolved form the surrounding soil. The atmosphere is a source of nitrogen, which is available as nitrates, nitrites or ammonia in water. When these nutrients are deficient, production and growth of organisms in water are affected.

Deficiencies is one or more of these important environmental factors have to be artificially made good in fish culture practices, as in agriculture.

To sum up, a good pond water for fish cultivation should have a slightly alkaline reaction (pH ranging from 7.5 to 8.5), a fair amount of dissolved oxygen of the order of 5 to 7 parts per million parts of water, a concentration of 0.2 to 0.4 parts per million of phosphates, and 0.06 to 0.1 parts per million of nitrates, and about 100 or more parts per million of total alkalinity in water. As the natural productivity of the pond, like any piece of cultivable land, will steadily decrease with the continued cultivation of fish in it, the maintenance of water conditions at the optimum level by periodic application of manures will be necessary. A regular chemical analysis of water samples from the pond to find out when and what corrective treatments are to be applied thus becomes essential.

The aquatic fauna, like the common plants found in fish ponds, can be classified as animal plankton (zooplankton) or free water organisms, weed-dwelling organisms and bottom-dwelling organisms.

Zooplankton

Important zooplankters usually found in India freshwater fish ponds belong to the Rotifera or wheel animalcules (*e.g., Brachionus, Filinia, Polyarthra, Pedalia, Keratella,* and *Asplanchna*) and the Crustacea dominated by the Cladocerans of water-fleas (*e.g., Daphnia, Moina, Ceriodaphnia,* and *Simocephalus*) and the Copepods-several species of *Diaptomus* and *Cyclops* and their young ones the Nauplius larvae. Ostracods are relatively less common.

The cladocerans and copepods, almost universally present, from the most important food items of small fish. Rotifers are also equally relished. In manured ponds, these often appear in dense swarms, but the last only for a few days.

Weed-Dwelling Fauna

Several rotifers, flatworms and water mites (hydracarina) which creep among the leaves of aquatic plants, the common Hydra, the

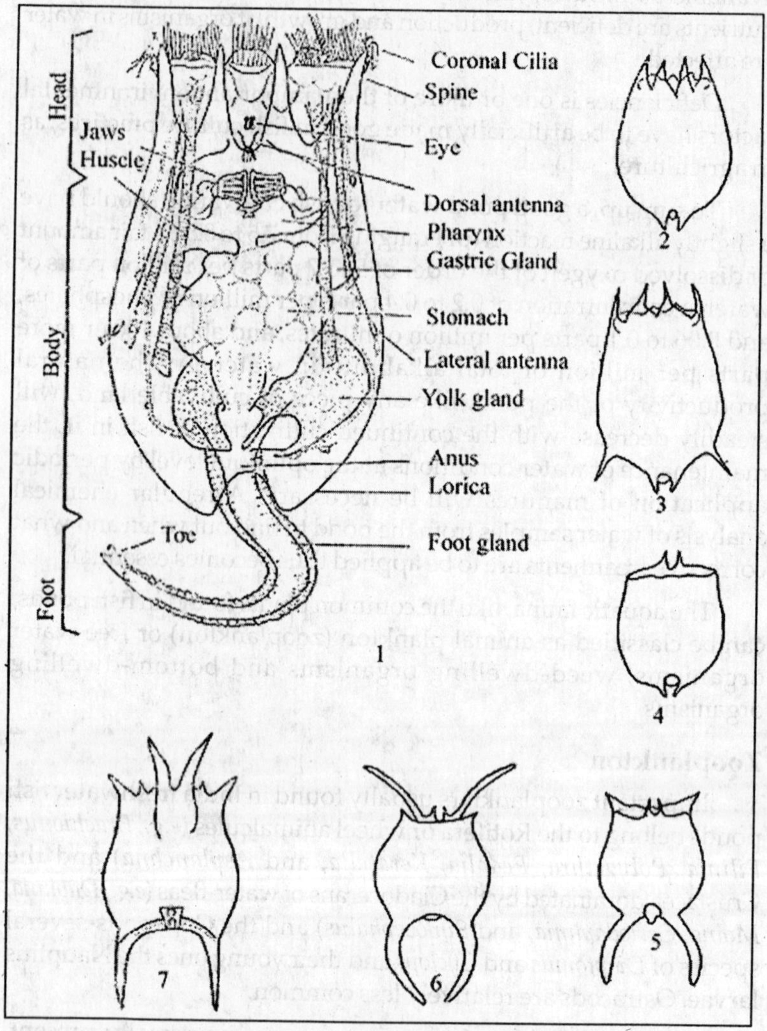

Figure 2.1: The Rotifers *Brachionus* sp. in Fish Pond

1. *Brachionus calyciflorus* (female-expanded); 2. *Brachionus calyciflorus* (without posterolateral spines); 3. *Brachionus calyciflorus* (with posterolateral spines); 4. *Brachionus angularis*; 5. *Brachionus caudatus*; 6. *Brachionus falcatus*; 7. *Brachionus quadridentata.*

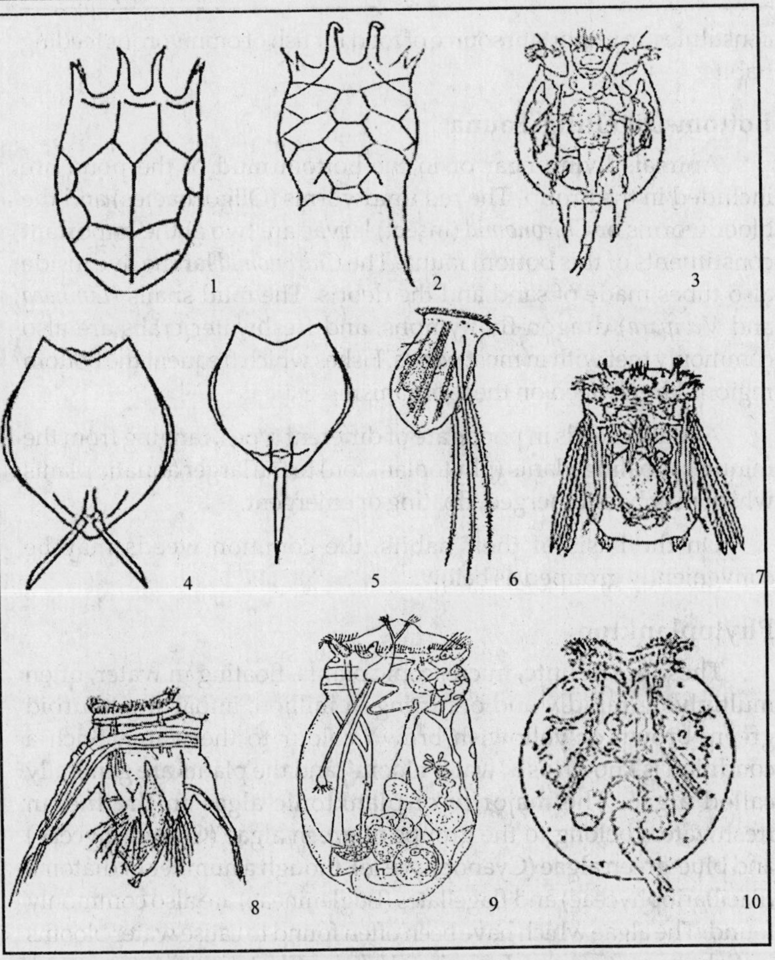

Figure 2.2: Common Rotifer in Fish Pond

1. *Keratella cochlearis;* 2. *Keratella tropica;* 3. *Platyias,* 4. *Lecane;* 5. *Monostyla;* 6. *Filinia longiseta;* 7. *Polyarthra;* 8. *Hexarthra,* 9. *Asplanchna;* 10. *Testudinella.*

snails, the bristleworms (Oligochaetes) and the numerous insect larvae are included under this category. Some of these, like the Hydra, are attached to the leaves and derive their subsistence from the deposits on the leaves, the small organisms that come near the leaves and also on the tender leaves of plants. This category of fauna constitutes an important source of food for fish of omnivorous feeding habits.

Bottom-Dwelling Fauna

Animals living near or in the bottom mud of the pond are included in the group. The red mudworms (Oligochaetes) and the bloodworms or *Chironomid* (insect) larvae are two of the important constituents of this bottom fauna. The *Chironomid* larvae live inside also tubes made of sand and the debris. The mud snails (*Limnaea* and *Vivipara*) dragon-fly nymphs, and freshwater crabs are also commonly met with in most ponds. Fishes which frequent the bottom regions largely feed on these organisms.

Aquatic weeds in ponds are of different types, ranging from the minutest floating plants (phytoplankton) to the larger aquatic plants which may be submerged, floating or emergent.

On the basis of their habits, the common weeds may be conveniently grouped as below.

Phytoplankton

These are minute, microscopic plants, floating in water, often multiplying rapidly and occurring in million, imparting a turbid green, brown or yellowish brown colour to the water. Such a condition is known as a 'water bloom', and the plants are generally called algae. The majority of planktonic algae in the Indian freshwaters belong to the groups of green algae (Chlorophyceae) and blue-green algae (Cyanophyceae), though a number of diatoms (Bacillariophyceae) and flagellates (Euglenineae) are also commonly found. The algae which have been often found to cause water blooms in fishery waters are *Eudorina, Volvox, Closterium, Actinastrum, Scenedesmus, Pediastrum, Microcystis, Anabaena, Oscillatoria, Euglena, Ceratium* and *Melosira*.

The microscopic plants are eaten as food by most of the higher animal organisms in water. Owing to their resistant cell wall, some of them, at least during certain stages, have been found to be unsuitable as food for small fish.

Filamentous Algae

These are ordinarily submerged in water though when growing in profusion they float up in masses. The common forms encountered

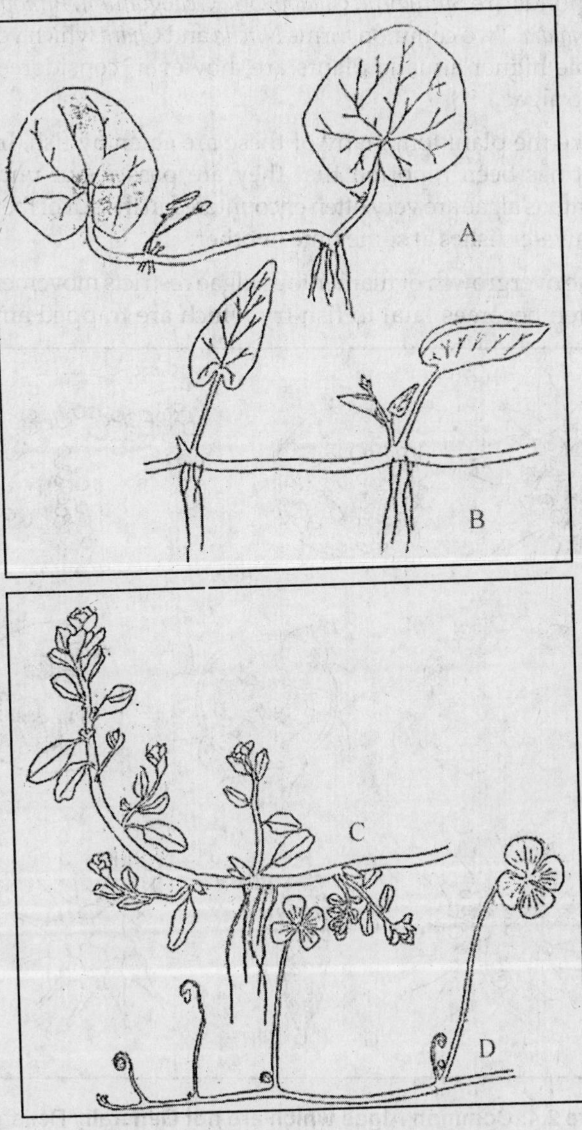

Figure 2.3: Common Aquatic Weeds

(A) *Limnanthemum cristatum*; (B) *Ipomea aquatica;* (C) *Fussiaea repens* and (D) *Marsilia quadrifolia*

in fish ponds are *Spirogyra, Cladophora, Oedogonium, Pithophora* and *Darparnaldia*. Two common forms *Nitella* and *Chara* which very much resemble higher aquatic plants are, however, considered closely allied to algae.

Like the plankton, many of these are eaten by fish. In certain cases it has been reported that they are passed out undigested. Filamentous algae are very often encountered in the gut of the majority of freshwater fishes at some time or other.

The overgrowth of filamentous algae restricts movement of fish and often becomes fatal to fish fry which are trapped among the

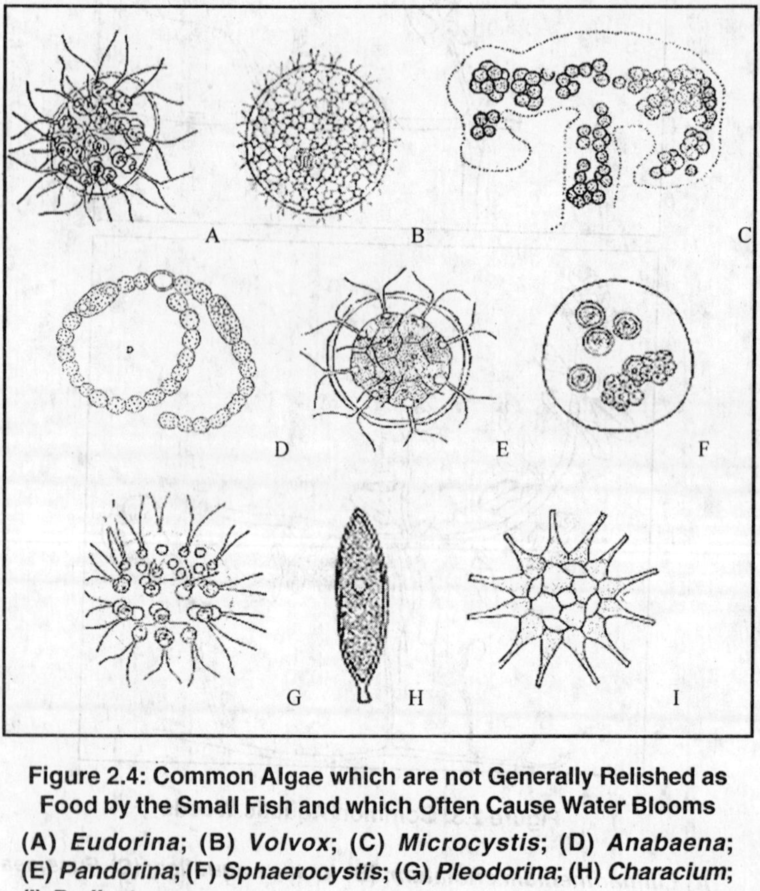

Figure 2.4: Common Algae which are not Generally Relished as Food by the Small Fish and which Often Cause Water Blooms

(A) *Eudorina*; (B) *Volvox*; (C) *Microcystis*; (D) *Anabaena*; (E) *Pandorina*; (F) *Sphaerocystis*; (G) *Pleodorina*; (H) *Characium*; (I) *Pediastrum*.

Contd...

Figure 2.4–Contd...

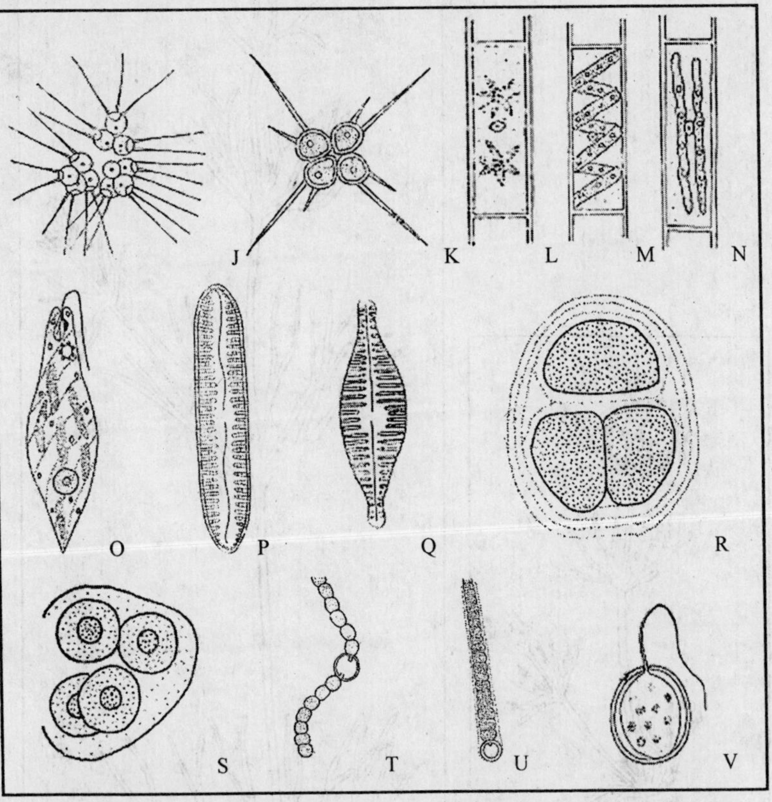

(J) *Micractinium*; (K) *Tetrastrum*; (L) *Zygnema*; (M) *Spirogyra*; (N) *Spirogonium*; (O) *Euglena*; (P) *Pinnularia*; (Q) *Navicula*; (R) *Chroococcus*; (S) *Gloeocapsa*; (T) *Nostoc*; (U) *Rivularia*; (V) *Trachelomonas*

filaments. Ordinarily, when these algae abound in water, plankton will be poor.

They are undesirable in nursery ponds and their over-growth is to be prevented in other waters.

Marginal and Emergent Weeds

These are rooted higher plants which abound in the shallow marginal areas of ponds and in swamps–the transitional zone between the terrestrial and the aquatic habitats.

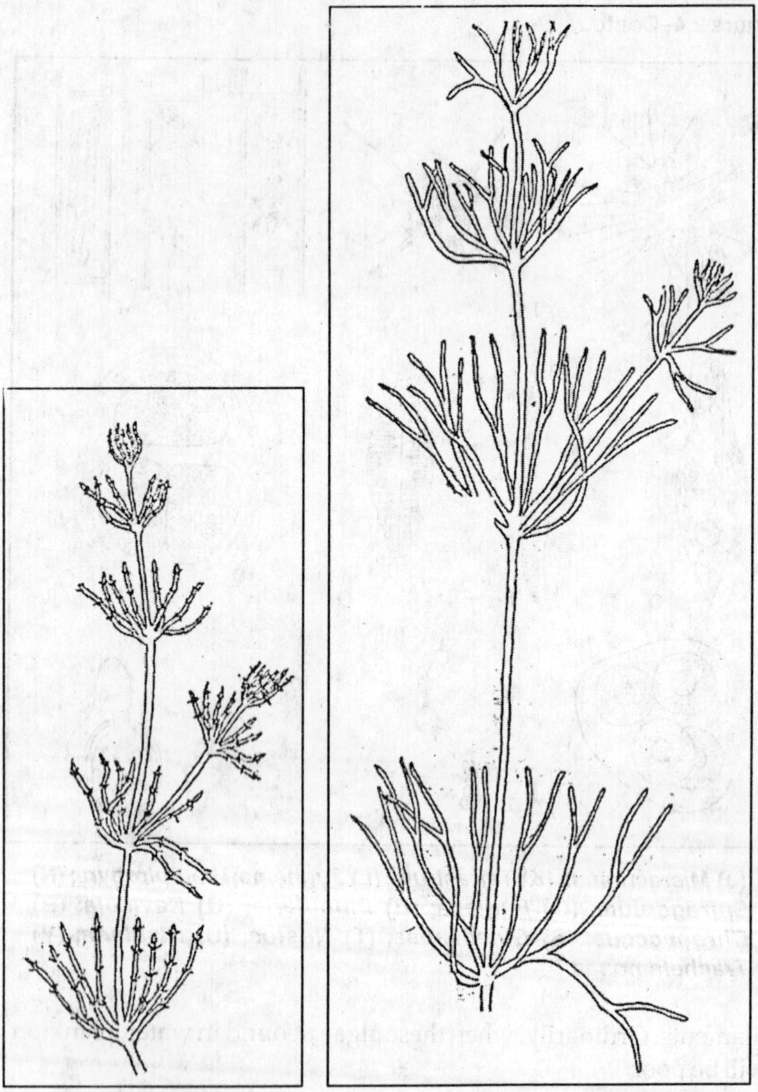

(A) *Chara* (B) *Nitella*

Figure 2.5: Common Submerged Aquatic Weeds

In nursery ponds they serve as shelter and as breeding spots for predatory aquatic insects and should, therefore, be carefully removed and kept in check.

Surface-Floating Weeds

The obnoxious water hyacinth (*Eichornia*) is a typical example. *Pistia* is another which often chokes up the water surface. The duck weeds *Lemna* and *Azolla* usually form green mats on the water surface.

Wolffia is one of the smaller duck weeds, small, green and grain-like, floating on the surface.

These often form a complete screen on the surface, shutting of sunlight and gases so that food production in the lower layers remains very poor. Overgrowth of these weeds is, therefore, to be prevented in fish ponds.

The duck weeds *Lemna* and *Wolffia* are reported to be used as food for the Chinese carps.

Submerged Plants

The majority of the submerged plants are rooted to the bottom. Typical examples are *Hydrilla, Otellia, Vallisneria, Potamogeton, Najas* and *Lagarosiphon*. These plants are ordinarily confined to the column of water and choke the surface only when there is excessive overgrowth during summer when the water level falls steadily and rapidly.

Some of the submerged plants like *Ceratophyllum* and the bladderwort *Utricularia*, have no typical roots, and, therefore, merely float in the water column. These plants are completely adapted to life in water and the latter (*Utricularia*) develops a series of characteristic round bladders, which trap minute aquatic animals to be used as food.

Other plants like the water lilies and the lotus (*Nymphea, Nelumbium* and *Limmanthemum*) are firmly rooted in the mud but their leaves often float at the water surface.

These true aquatic weeds often grow in such profusion that within a few weeks time the whole pond gets choked, the water is impoverished of nutrients and the growth of plankton is thereby hampered. A few fishes like the *Gourami* and known to make use of these plants as food. In the case of others, these hamper their movements and badly affect their food supplies. Fishing also becomes difficult in such ponds.

When weeds abound, wide fluctuations in water conditions occur, which affect the well-being of the fish.

Some of these weeds, at least in the rotting conditions, are eaten by carps, but whether this is actually of any appreciable nutritive value for the fish is not still known. Observations show that of the major carps, Catla does not eat submerged weeds to any appreciable extent, though Rohu and Mrigal probably utilize them to a limited extent.

Aquatic Weed Control

It is not ordinarily possible to control the algae or weeds by chemical agents without adverse effect on the animal populations in water. Copper sulphate is one of the well-known algicides used for controlling algae in filter beds in water works. As it is poisonous to fish and zooplankton, it cannot be employed in nursery ponds. However, application of copper sulphate at a daily dose of 0.3 parts per million (ppm) parts of water for several days has been reported to have given satisfactory results in the control of filamentous algae, without affecting the large fish.

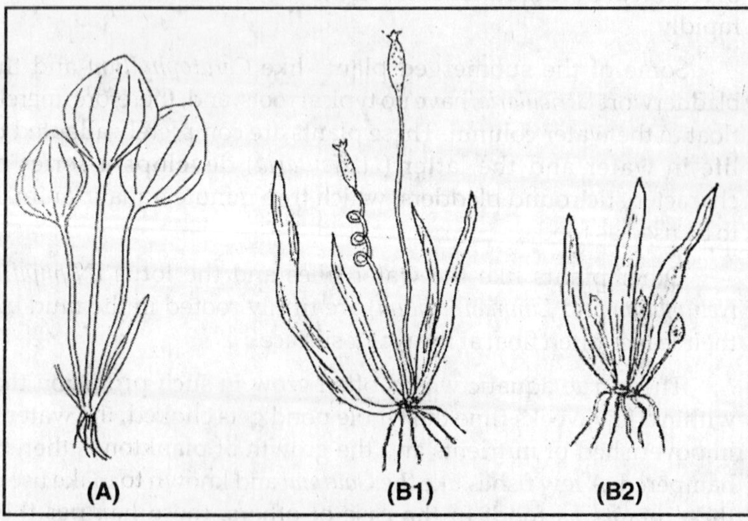

Figure 2.6: Common Submerged Rooted Aquatic Weeds

(A) *Otellia alismoides*; (B) *Vallisneria spiralis* (1) Female plant, (2) Male Plant.

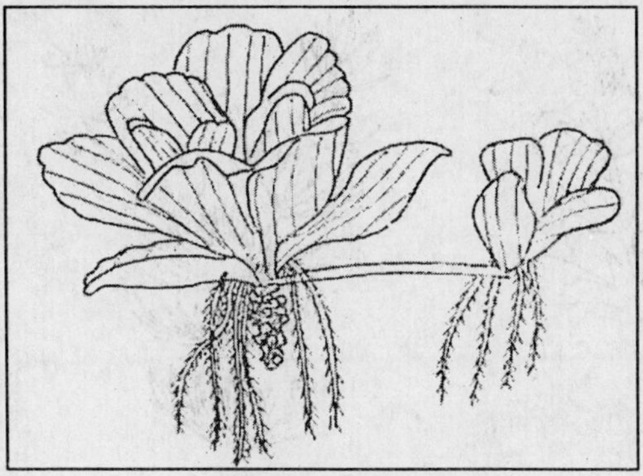

Pistia stratiotes

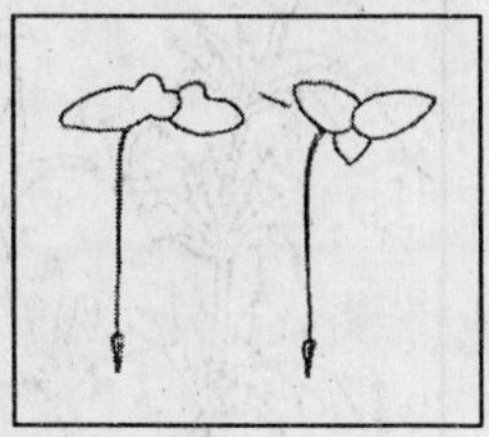

Lemna minor

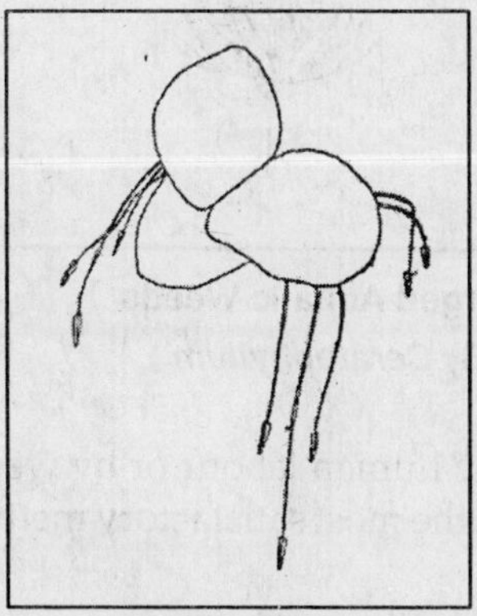

Lemna polyrrhiza

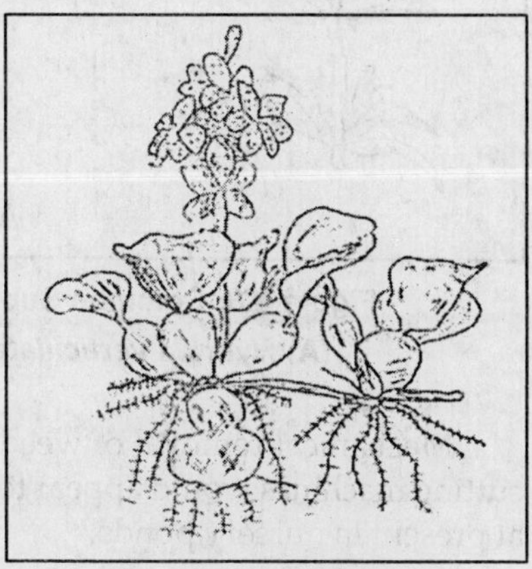

Eichhornia crassipes

Azolla pinnata

Figure 2.7: Common Floating Aquatic Weeds

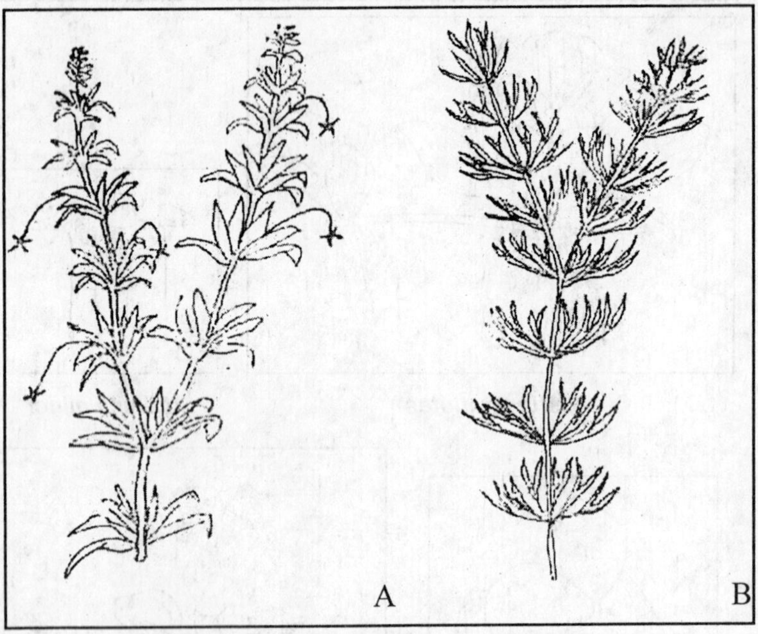

Figure 2.8: Common Submerged Aquatic Weeds (A) *Hydrilla verticillata*; (B) *Ceratophyllum*

Mechanical removal of weeds by human labour or by weed-cutting machines would appear to be the most satisfactory method at present in nursery ponds.

When labour is cheap, mechanical removal of weeds can be attempted in larger waters also. Once the weeds are thus removed, a careful watch and periodical removal of any new sprouting plants would effectively keep them in check.

Hormone weedicides like Methoxone, Agroxone and 2, 4-D have all been reported to be effective under certain conditions. "Dicotox", containing the ethyl ester of 2, 4-D, at a dose of one ounce per 100 square feet has been reported to effectively destroy submerged weeds like *Hydrilla, Najas, Vallisneria* and *Nymphea*, in about 20 days. When the dose is doubled, the plants are killed in about 16 days. With either of the above dose no harmful effect has been observed on fish or other animal life in the water.

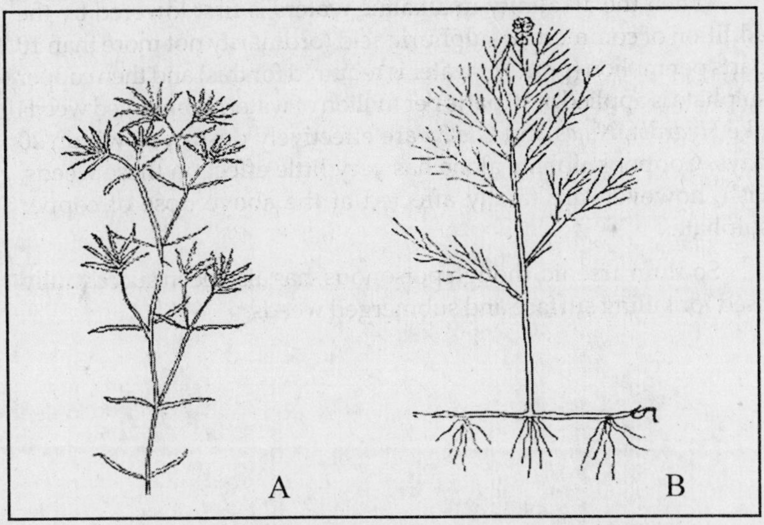

Figure 2.9: A Common Submerged Aquatic Weed
(A) *Najas minor*; (B) *Potamogeton pectinatus*

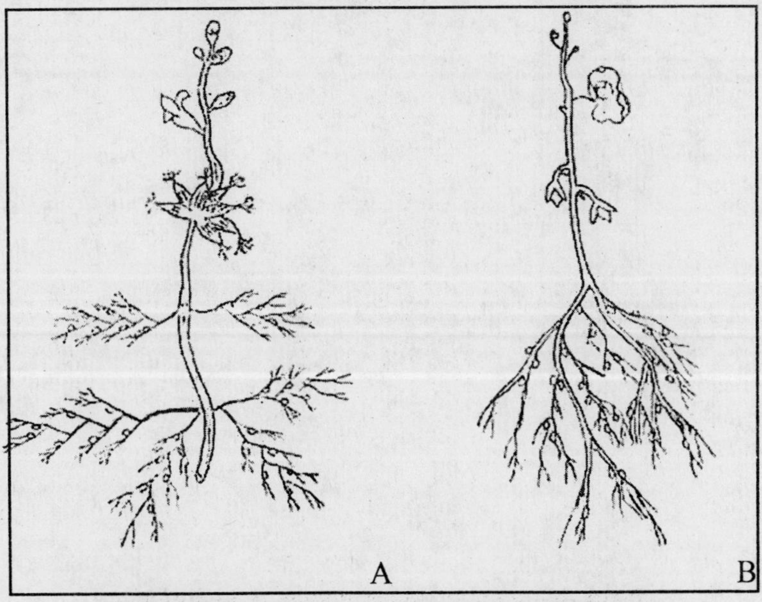

Figure 2.10: Common Submerged Aquatic Plants

(A) *Utricularia stellaris* (Plant with bladders, leaf segments and whorled floats); (B) *Utricularia flexuosa* (plant with bladders flowers and fruits)

When the alkalinity in alkaline waters is first lowered by the addition of commercial sulphuric acid (ordinarily not more than 10 parts per million (ppm.) of water is required for this) and then copper sulphate is applied at 10 parts per million of water, submerged weeds like *Hydrilla, Najas* and *Otellia* are effectively destroyed within 20 days. Copper sulphate alone has very little effect on these weeds. Fish, however, are fatally affected at the above dose of copper sulphate.

Sodium arsenic, though poisonous, has also been successfully used for killing surface and submerged weeds.

Chapter 3

Biology of Cultured Carps

A large variety of suitable indigenous species of fishes are available in various parts of India for cultivation in ponds. Some exotic species of food fishes have also been introduced into this country. The cultural value of these fishes varies considerably, depending upon their relative growth, size, adaptability to different types of waters, and easy availability.

The large majority of these fishes belong to the group of carps which are utilized for culture. Murrels, a few species of catfishes and perches, and a few quick-growing salt water fishes are also commonly utilized of cultivation in brackish water and freshwater ponds.

Selection of Cultivable Species

Selection of species of fish, which grow rapidly and attain a large size, besides being omnivore feeder and breeding under any conditions and such fishes which have good market value is essential for fish culture. For transplantation, such species need be hard enough to enable transportation of young fry and fingerlings, over long distances.

Fishes which ordinarily feed on microscopic plants (Phytoplankton) and animals (Zooplankton) in water, on the decaying weeds and debris are much more efficient converters of organic matter, into fish flesh, than those species, which subsist by preying upon other fishes.

The former fishes belonging to Indian major carps group are therefore preferred for Aquaculture. The most important of these fishes are : catla (*Catla catla*), rohu (*Labeo rohita*), calbasu (*Labeo calbasu*) and mrigal (*Cirrhinus mrigala*). These species are commonly found in the Ganga, Brahamputra and Mahanadi river systems.

Catla and Mrigal are found in Narmada and Godavari rivers also. During recent past these have been successfully introduced into the South Indian rivers, where they have established.

The Indian major carps fulfill the qualities of culturable fishes, except that these do not breed in confined waters. To breed them in confined waters, therefore, the method of breeding by hypophysation is adopted which is highly successful in most of the normally prevalent field conditions during south-west monsoon.

Besides Indian major carps, some exotic carps have also been imported, which exhibit good growth as Indian major carps, when these are cultured along with Indian major carps. These carps are Grass carp (*Ctenopharyngodon idella*), Silver carp (*Hypophthalmichthys molitrix*) and Common carp (*Cyprinus carpio*). Of these exotic carps only *Cyprinus carpio* breeds in confined waters without hypophysation. The other two exotic carps are induced bred by stimulating with suitable dozes of pituitary hormone injections.

The purpose of importing exotic fishes is to fill up the gap in ecological niches, left by indigenous fishes or to culture them individually in waters unsuitable for indigenous fishes.

As regards to ecological niches, fertile tropical zone containing micro- and macro-vegetation from niche which is utilized by indigenous carps. But unless all space and food zone in a water body is judiciously utilized maximum production may not be attained. It is for this reason that the composite fish culture consisting of Indian major carps and exotic carps in different species composition is adopted.

Before taking up the culture of the Indian major carps (IMC) and exotic carps, it is necessary to know their biology. Biology of the carps will reveal the methods to be adopted in the management in regard to their feeding, breeding and rearing. Biology of IMC and exotic carps are therefore described below.

Biology of Cultured Carps

The species of fish dealt with, either individually or as a group, are commonly called carps. The species referred to are primarily seven in number and are often grouped on the basis of their natural geographic occurrence :

The so-called Chinese carps, which include the grass carp, *Ctenopharyngodon idella*; the silver carp, *Hypophthalmichthy molitrix* and the exotic bighead carp, *Aristichthys nobilis* and the so-called Indian major carps, which include catla, (*Catla catla*), rohu (*Labeo rohita*) and mrigal (*Cirrhinus mrigala*). The seventh species is the common carp, *Cyprinus carpio.* Each of these species by common parlance, is known as carp. Ordinarily, however, the expression "the carp" or "carp", when used in singular for an individual fish, generally can notes the common carp, *Cyprinus carpio*, rather than any other cyprinid species. Other fishes known by their own individual particular names like silver carp, grass carp, catla, rohu etc. Taxonomically, carps belong to the family Cyprinidae.

Following is a brief description of the significant biological characteristics relevant to aquaculture *viz.*, geographical distribution, food, growth, size and age at first maturity and fecundity, especially important from the point of view of operating a hatchery, of each of the species of carps dealt here in this book.

Catla (*Catla catla*)

Distribution

Catla is the natural inhabitant of the freshwater sections of the rivers of northern India, Pakistan, Bangladesh and Burma.

Habitat

The favourite habitat is the deep pools of the rivers of North India. The species has been transplanted into some of the rivers of peninsular India, notably River Cauvery and in more recent times, into Sri Lanka and China.

Characters

Large upturned mouth, barbells absent, dorsal origin before pectoral fin, body deep, moderately compressed, head broad, mouth

Figure 3.1: Catla (*Catla catla*)

wide, lower jaw prominent, caudal fin deeply forked, length of head 4/ 4 ¼–4 ¾ of caudal 10. All fins rather elongated.

Commonly known as Catla. It is the fastest growing Indian major carp (IMC) of India.

Food and Feeding Habits

The dominant occurrence of zooplankton in the gut contents of catla indicates that it is mainly a water column feeder. The incidence of organic detritus, mingled with sand, mud and rooted aquatic plants indicates bottom browsing habit as well. The presence of certain unattached submerged floating vegetation points out that the fish also explores the mid-layers of water.

Growth

Catla is the fastest growing of the Indian major carps of India. In natural waters, catla attains a length of 295 mm (weight 354 g) in the first year, 514 mm (weight 2,193 g) in the second year, 716 mm (weight 6,501 g) in the third year, 823 mm (10,282 g) in the fourth year and 917 mm (weight 14,665 g) in the fifth year. It attains 6 feet in length (1830 mm.), and weight more than 60 Kg (Day 1878). The growth varies according to ecological factors.

Maturity

Catla attains first maturity in the second year of life.

Fecundity

The fecundity of catla vary from 230,831 to 4,202,250 depending upon the length and weight of fish and the weight of the ovary.

Spawning Season

The spawning season of catla coincides with the Southwest monsoon in North-eastern India and Bangladesh, where it lasts from May to August and in North India and Pakistan, from June to September.

After spawning, the ovaries are in spent condition, flaccid and bloodshot, with a few dead detached ova lying in the lumen of the ovary.

A fully swollen, water hardened egg of catla has a volume of approximately 0.08 ml. A catla hatchling after its yolk sac is absorbed weighs approximately 0.0025 g.

Rohu (*Labeo rohita*)

Commonly known as Rohu or sometimes as Rohi. In growth it is next to Catla amongst the Indian major carps (IMC).

Distribution

Rohu is the natural inhabitant of freshwater sections of the rivers of North India; the River Narmada, Tapti and Mahanadi in central India and rivers of Pakistan, Bangladesh, Burma and the Terai region of Nepal.

It has been transplanted into some of the rivers of peninsular India and Powai Lake, Mumbai. Rohu has also been transplanted to Sri Lanka and to Mauritius.

Characters

Dorsal profile more arched and slightly curved, mouth inferior, snout obtuse, lips thick and fringed, dorsal origin is midway between snout and caudal base, caudal deeply forked.

Habitat

Rohu is the natural inhabitant of freshwater sections of the rivers of North India. It was later transplanted into Southern Indian

Figure 3.2: Rohu (*Labeo rohita*)

rivers, where it has established. It is found in freshwaters of Pakistan, Bangladesh and Burma and has been transplanted in many other countries. Although freshwater fish it can live in slightly brackish water.

Quick growing carp, it is largely used for culture in ponds all over the country. It is esteemed food fish of West Bengal, Bihar and Uttar Pradesh, where it forms an important item in composite fish culture.

Food and Feeding

Rohu is a bottom and column feeder and prefers to feed on plant matter including decaying vegetation; it is less adapted to take zooplankton than even mrigal. Utilization of plant matter is much better in mrigal and rohu than in catla. Rohu fingerlings 100–250 mm long subsist on unicellular and filamentous algare (15 per cent), rotten vegetation (55 per cent), rotifers and protozoans (2 per cent) and crustaceans (8 per cent). The rotten vegetation component in the food increases in bigger fish.

Growth

Under culture conditions, highly variable growth has been mentioned by different workers. It is a very quick growing fish though having a somewhat slower growth rate than catla.

Maturity

Rohu attains sexual maturity towards the end of the second year in ponds. The gonads started developing during the month of February, the ripe stage was reached during June and July and finally spent individuals were found in late July and early August. Temperature range is from 27° C to 30° C (28.5° C optimum) with rain water rich in oxygen. Environment not needed for maturity but needed for shedding of eggs, which are influenced by hormones secreted by pituitary gland. Pattern of distribution is 2 males to 1 female. Fertilization is external.

During the spawning season, males were found to mature earlier than females. The larger-sized individuals were also found to mature earlier than the smaller ones. The minimum age at first maturity for both sexes was two years while complete maturity was reached by males at age four years and by females at age five. The 50 per cent maturity point was 2.7 years for males and 2.9 years for females. The minimum size at first maturity of males was 46.2 cm and 100 per cent maturity was found in the 650 mm length group. The smallest mature female observed was 515 mm and all the females were mature in the 700 mm length group. The 50 per cent maturity points were 549.0 mm and 579.6 mm for males and females, respectively.

Fecundity

The fecundity of rohu is reported to vary from 226,000 to 2,794,000 depending upon the length and weight of the fish and weight of the ovary.

Spawning Season

The spawning season of rohu generally coincides with the southwest monsoon, though it appears to be somewhat variable in different parts of India as given below :

Spawning Season	*Locality*
July-August	Western Uttar Pradesh & Punjab
June-August	Orissa & West Bengal
April-August	Ganga system of North Bihar
July-August	Ganga system (Aligarh, Uttar Pradesh)
April-May-June	Assam
June-September	Aligarh

Eggs are slightly reddish transparent, non-adhesive and demersal. Eggs measures 1.5 mm, swells up on fertilization and measures 1.5 mm. Hatching takes place in 16 – 18 hours at optimum temperature of 27º C. Hatchlings measure 3.8 mm. A fully swollen, water-hardened egg of rohu has a volume of approximately 0.078 ml. Narrow, portion of yolk equal or slightly greater than bulbous part. A rohu hatchling after absorption of its yolksac weighs approximately 0.0021 g.

Mrigal (*Cirrhinus mrigala*)

Distribution

Mrigal is the natural inhabitant of the freshwater sections of the rivers of Northern India, Bangladesh, Burma and Pakistan. It has been transplanted into waters of peninsular India for aquaculture.

Habit and Habitat

Mrigal is a detritus eater with a narrow range of food variety. It is a bottom feeder subsisting mainly on decayed vegetation. Some scientists consider mrigal an omnivore also frequenting the water column for feeding.

Food and Feeding

Semi-decayed organic matter constitutes about 65 to 78 per cent of the gut contents of the different size groups of mrigal, with semi-digested organic matter followed by plankton predominating in the

Figure 3.3: Mrigal (*Cirrhinus mrigala*)

size groups 561 mm to above 766 mm fish length and the reverse in the younger age groups measuring upto 560 mm length considerable amounts of sand and mud, the former measuring up to about 21 per cent of the gut contents and the latter upto about 13 per cent are encountered.

Growth

In natural waters, the fish shows a very rapid growth rate in the first years of its life, followed by a period of slow growth in the next three years. The growth rate thereafter becomes even slower.

Under conditions of aquaculture, the growth rate is a function of stocking rate, natural food available and feeds given, competition with other co-stocked fish and environmental conditions.

Maturity

Mrigal is reported to attain its first maturity when about one year old or two years old. The males mature at the end of the first year and females, some time later.

Fecundity

The fecundity of mrigal to vary from 124,800 to 1,905,000 in specimens weighing 904 g and 9,503 g, respectively.

Mrigal contains a single group of maturing eggs in the ovaries and spawns once a year. The size of oocytes started to increase in April. The maximum size of ova was recorded during June. Spent fish contained either few or no mature ova in the ovary during post-spawning months. The size of mature ova varied from 0.92 mm to 1.10 mm.

Spawning Season

The spawning season of mrigal depends on the onset and duration of the monsoon. It coincides with the Southwest monsoon in India, Bangladesh and Pakistan. The duration of the spawning season varies in different regions of the subcontinent as shown below:

Spawning Season	*Locality*
July and August	Punjab
June and July	West Bengal
April–August	Ganga River System (depending upon floods in each region);

Spawning Season	Locality
June – August	Western U.P., Northern India (River Ganga)
July–September	South India (The Godavari and Krishna)

Eggs

A fully swollen and water hardened laid egg of mrigal has a volume of approximately 0.11 ml. A mrigal hatchling after its yolksac is absorbed weighs about 0.0025 g.

Calbasu (*Labeo calbasu*)

Commonly known as kalabans or kalhans. It is next to mrigal in growth.

Characters

Dorsal and ventral profiles almost equally convex. Colour dark. Body deep, Mouth rather narrow, small, tapering head. Snout obtuse. Lips thick and fringed. Pores on snout and upper lip. Chins slopes away from mouth. Mouth is subterminal. Two pairs of fleshy barbells, rostral larger than maxillary. Dorsal midway between snout and caudal base.

Growth

It attains a maximum size of four feet (1916 mm). In the 1st year the growth is 30 – 45 cms. With 484 gms weight. When stocked with other carps the growth is 450 gms in the 1st year.

Food and Feeding

In early stages the fry feeds on zooplankton or unicellular algae. As it increases in size the food contents contain 30 per cent phytoplankton, 50 per cent zooplankton (Rotifers & Crustaceans) and the remaining food consists of organic matter and debris, detritus and benthic fauna.

Maturation

Matures at 2 years in life, when 21 – 23 cms in length.

Fecundity

Specimens of 500 and 1000 gms contain 438 and 288 eggs per gram, respectively.

Spawning Season

In rivers from July to September. Spawning once a year. Exhibits local migrations for spawning.

Environment and Spawning Behaviour

Temperature range is from 27° C to 30° C (28.5° C optimum) with rain water rich in oxygen, movements of water in marginal areas of rivers. Environment not needed for maturity but needed for shedding of eggs which are influenced by hormones secreted by pituitary gland which is first to receive external stimuli. Pattern of distribution is 2 males to 1 female. Fertilization is external.

Eggs and Hatchlings

Eggs are transparent, demersal and non-adhesive, oval and bluish in colour. Swell on fertilization and hatch out in 16–18 hours at 27° – 29° C as optimum temperature. Hatchlings attain adult characters in 30 days when they reach to fry stage.

Production

Not cultured because of slow growth.

Distribution

Occurs in freshwaters, almost throughout India, Pakistan, Bangladesh and Burma.

Economic Importance

The fish is not available in large number. Whenever, available the fry is stocked in ponds and tanks but being slow in growth, it is not preferred in fish culture. It is valued as food fish and as game fish.

Grass Carp (*Ctenopharyngodon idella*)

Introduction

Commonly known as grass carp and also referred to as white Amur, the fish has been transplanted to various parts of the world for weed control. It is a natural inhabitant of the flatland rivers of China and the middle and lower reaches of river Amur in the former USSR and the fish has been introduced into many other countries in recent times. In some of the countries the main purpose of its introduction, in addition to culture, is biological aquatic weed control in natural waterways, lakes and man-made lakes.

In India it was introduced in the year 1959 at the Central Inland Fisheries Research Institute (ICAR), Barrackpore, Sub-station, Cuttack from Hongkong. Another consignment was brought in 1961.

Characters

In colour it is dark grey above and silvery on the belly. Body elongated and moderately compressed. Head broad with short round snout, upper jaw slightly longer than lower jaw. No barbells. Short dorsal fin rays (10). Dorsal origin slightly in advance or opposite to that of ventral. Scales of moderate size. Grass carp, like other cyprinids, has a toothless mouth but has specialized pharyngeal teeth for rasping aquatic vegetation. These teeth occur in two rows, the upper consisting of two small teeth on either side and the lower of strong comb or file-like teeth comprising four on the right and five on the left pharyngeal bone. In fishes of total length 30 cm and below, the lower pharyngeal teeth have a serrated cutting surface, while in larger fish, the teeth are thicker and have double flattened serrated cutting and rasping surfaces. These patterns of teeth structure are associated with the change in the feeding habit. Large fishes are able to masticate the leaves of tough land plants such as fibrous grasses.

Digestion in the grass carp is said to be incomplete and about half the food material ingested is excreted as faeces which, it is believed can support directly or indirectly a large biomass of other species of fish.

For fish above 30 mm, the natural diet is virtually exclusively macrovegetation, tender aquatic weeds like *Wolffia* initially and a wide variety of other aquatic macrovegetation and even softer land plants as the fish grows. Among the macrovegetation, the preferred weeds in the different regions of the world are : *Wolffia, Lemna, Spirodela, Hydrilla, Najas, Ceratophyllum, Chara, Potamogeton, Vallisneria* and *Myriophyllum*.

Growth

Growth varies in different environments and in different countries. In India with a stocking rate 3500 fish/ha ordinarily it grows to 2200 mm and 20 Kg by weight but growth varies with density of stocking as availability of weeds is primary factor in its growth.

In India fingerlings when stocked at the rate of 300 – 500 fish/ha in polyculture gives a daily growth of 4.7 gms. In natural waters, grass carp attains a length of 15 to 30 cm weighing 225 g to 650 g at the end of the first year; a length of 60 cm and a weight of 1.8 to 2.3 kg at the end of second year. After four years, the weight may be 4.5 kg.

Under culture conditions, growth is a function of stocking rate and feeds given and their conversion rate, competition with other fish co-stocked with grass carp and environmental conditions.

Food and Feeding

Early fry to late fry feed on mainly zooplankton with occasional food of phytoplankton, and slightly decomposed higher plants and soft parts of stems and roots. From fingerling to adult stage it feeds on aquatic macrovegetation, demersal *Hydrilla* and *Ceratophyllum*, floating *Lemna* and *Wolffia* and soft parts of plants. The fish has been observed to clear off *Vallisneria, Ottelia, Utricularia, Trapa, Myriophyllum* etc. and other aquatic weeds such as *Potamogeton, Nitella* and filamentous algae like *Spirogyra* and *Spethophora*. The fish does not appear to feed actively on *Eichhornia, Pistia* and *Nymphia* which can not be controlled by this fish.

Digestion in the grass carp is said to be incomplete and about half the food material ingested is excreted as feces which, it is believed, can support directly or indirectly a large biomass of other species of fish.

Maturation

In temperate regions it matures from 2nd to 3rd year, in some countries when weight of fish is 2 Kg about or higher but in very cold countries it gets matured during 5 – 6 years age when its weight is 2.5 Kg. In India all induced bred stock mostly matures within zero year classes. Males maturing a month or so earlier and the rest in 1 year growth. Age at which grass carp attains maturity varies greatly with climate and environmental factors, especially temperature.

Fecundity

Fecundity varies from 309000 to 618000 eggs in fishes weighing from 4 – 9 Kg (74 cms long) to 7.0 Kg (79 cms), 52000 to 134000 eggs per Kg of body weight have been recorded. Fish weighing 4,766 to 7,036 g had 308,000 to 618,100 ova, about 82 ova per gram body weight of the fish and 610 ova per gram weight of the ovary. The ovarian eggs of grass carp have a diameter varying between 1.19 and 1.37 cm. A fully swollen water-hardened spawned egg of grass carp has a volume of approximately 0.055 ml. A grass carp hatchling after absorption of the yolk sac weighs approximately 0.0022 g.

Spawning

Breeds only in lotic environment, during Southwest monsoons from June to September and once in a year. In India it has been successfully bred by hypophysation. Eggs are copper colour and hatching period is 18 – 20 hours. The fish breeds during monsoon months in the flowing waters of its natural habitat, the rivers, but does not spawn naturally in the static waters of ponds and tanks.

Distribution

Autochthonous in freshwaters of China and Russia, it is native of South, Central and North China and middle and lower sections of river Amur in USSR. It has been transplanted and cultured in large number of countries of temperate regions and lesser number of countries than silver carp in tropical regions, of the world, including India.

In many countries like Japan and Taiwan it has been well established. In India it was introduced in 1959 at Cuttack substation from Hong Kong and later in 1961 from the same place to the same place. It was successfully bred by induced breeding in India in the year 1962.

Economic Importance

Phytophagus, it is widely cultured in temperate subtropical and tropical regions for biological aquatic weed control. Essential component of polyculture, the fish occupies a separate niche and does not compete with other fishes. Being aquatic weed feeder, the undigested food excreted by the fish forms food of other fishes besides it mineralizes water to make available more nutrients for emanating primary production which sustains secondary production or zooplankton. Phytoplankton forms food of silver carp and zooplankton that of *catla* in composite fish culture besides forming food of other fishes. Induced bred in many states it is hardy with wide range of fluctuations, specially it tolerates salinity upto 7.5 to 8.5 ppt from pre-fingerlings stage onwards and wide range of temperature too and in this aspect it is better tolerant than Indian major carps and Silver carp.

The fish exhibit high conversion ratio 12 Kg fish added in a pond filled with *Ceratophyllum* exhibited a growth of 6 Kg in 6 months. For aquatic weed control the requirement is 100 Kg/ha.

Rotten vegetable falling during browsing becomes feed for bottom feeders. Grass carp works wonders in clearing aquatic weeds in a fantastic manner exhibiting fantastic growth. It is highly useful for eradicating submerged aquatic weeds. Marginal aquatic plants are also consumed. Floating aquatic weeds are also consumed by young ones. Water hyacinth, *Pistia* and *Salvinia* are not controlled by grass carp. For aquatic weed control 1 Kg at the rate of 100 fish per hectare are introduced as in 6 – 8 months pharyngeal teeth of the fish become strong enough to feed on aquatic weeds.

Grass carp, like other cyprinids, has a toothless mouth but has specialized pharyngeal teeth for rasping aquatic vegetation. These teeth occur in two rows, the upper consisting of two small teeth on either side and the lower of strong comb or file-like teeth comprising four on the right and five on the left pharyngeal bone. In fishes of total length 30 cm and below, the lower pharyngeal teeth have a serrated cutting surface, while in larger fish, the teeth are thicker and have double flattened serrated cutting and rasping surfaces. These patterns of teeth structure are associated with the change in the feeding habit. Large fishes are able to masticate the leaves of tough land plants such as fibrous grasses.

Silver Carp (*Hypophthalmichthys molitrix*)

Popularly known as silver carp. It was introduced into India in the year 1959, at substation Cuttack (Orissa) of the Central Inland Fisheries Research Institute (ICAR), Barrackpore (West Bengal), from Japan.

Characters

Body oblong, slightly compressed, head pointed, snout bluntly rounded, lower jaw slightly protruding, small eyes, very small scales. Origin of dorsal behind ventral, midway between tip of snout and base of caudal. Abdomen keeled from vent to thorax. Dark grey above and silvery at the belly.

Silver carp naturally occurs in the river systems, Yangtze, West River, Kwangsi and Kwantung in South and Central China and in the Amur Basin in USSR and the species has been introduced into many countries in recent years for aquaculture.

The fish shows certain anatomical and morphological modifications in correlation with its phytoplanktophagous feeding habit. The length of the gut of the adult fish is 15 times the body length. The gills of silver carp have a complex network and profusion of closely set gill-rakers.

Growth

It is fastest growing carp and at optimum stocking rate forms 40 per cent of total catches in composite culture and grows faster in lotic (running) waters than in lentic (confined) waters. There is little information available on the growth rate of silver carp in rivers of its natural occurrence. Under culture conditions, growth rate is a function of stocking rate, natural food available and feeds given, competition with other species in polyculture, conversion rate of feed and environmental conditions. Under a given set of conditions growth rate of fry of silver carp is extremely high in the first 10 days the fish doubling its weight every second day and becoming about 19 mm long weighing 0.09 g in 10 days, 47 mm long weighing 1.1 gm in 20 days and 17 cm long weighing 5.5 g in 60 days. Absolute weight increases 0.001 – 0.02 g/day in the first 10 days and 4.2 g/day during the fingerling stage. Silver carp attains higher growth rate in length in the second year of life and maximum growth rate in weight in the third year. Growth in both length and weight declines sharply after the third year, by which period the fish may weigh as much as 2,780 g, gaining weight at the rate of 6.3 g/day.

Food and Feeding

In spawn stages it feeds on nannoplankton shifting to zooplankton including Protozoa, Rotifera and Naupli and others.

From early fry to late fry stage it feeds on zooplankton as main food with phytoplankton as occasional food but later on phytoplankton becomes the major food with zooplankton as occasional food. Phytoplanktivorous surface feeder from fingerling to adult stage, its main food is euglenoids, *Microcystes, Nostoc, Diatoms, Desmids*, filamentous green algae. Throughout life with occasional food of zooplankton. Micro-phyto planktophagic species remains throughout its life as pelagophagic.

Maturation

In temperate countries it matures in 2 – 3 years of its life. In some countries when weight of fish is 2 Kg about or higher; but in very cold countries during 5 – 6 years when weight is 2.5 Kg. In India in original stock at Cuttack the male developed oozing condition in years whereas female also matured in 2nd year.

All induced bred stock in India mostly matures within zero year class, male maturing a month or so earlier and the rest in 1 year growth. Of the environmental factors, temperature exercises maximum effect on the maturity of silver carp.

Fecundity

1.45 lakhs to 20.44 lakhs of eggs have been observed in fishes from 3 – 8 Kg average fecundity being 170,000 eggs per Kg of body weight. The fecundity of silver carp weighing 3.18 kg to 8.51 kg, as 145,000 to 2,044,000. The number of eggs per g body weight was 171 and per g ovary weight, 292.

Spawning

Spawns once a year by hypophysation from June to August. Eggs are dirty white, demersal and non adhesive. Measure 1.2 mm to 1.4 mm but swell up after fertilization when measure 4.9 mm.

Distribution

Autochthonous in freshwaters of China and Russia, transplanted and cultured in large number of countries of temperate and subtropical regions of world including India.

The fish has established in south east countries like Indonesia, Taiwan, Hongkong and Philippines and in other countries of the world. In India it was introduced at Cuttack substation of the Central Inland Fisheries Research Institute (ICAR), Barrackpore (West Bengal) from Japan in 1959.

The eggs from the yearlings to silver carp, 0.6 to 2.4 kg in weight, measured, on average, 1.20 mm in diameter, whereas, those of 4 year old fish were about 11 per cent larger. A fully swollen, water-hardened silver carp egg has a volume of about 0.07 ml. A silver carp hatchling after yolk sac is absorbed weighs approximately 0.0031 g.

The fish breeds naturally during April – July in the flowing waters of its natural habitat, the rivers of China.

Economic Importance

Microphyte planktophagic, it is now fastest growing carp of India. Highly useful in composite fish culture, where it slightly competes with catla. Extremely useful for stocking reservoirs and tanks rich in phytoplankton

Common Carp (*Cyprinus carpio*)

The fish species has 3 varieties which are the:

1. *Cyprinus carpio* var. *communis* (German strain)
   Cyprinus carpio var. *communis* (Bangkok strain)
   Commonly known as the Scale carp.
2. *Cyprinus carpio* var. *specularis* (Mirror carp)
3. *Cyprinus carpio* var. *nudus* (Leather carp)

A pond breeder of the aforesaid species was introduced into India to meet the requirement of fish seed. German strain of mirror carp introduced in Ooty lake in 1939 from Ceylon which did not work well in warm waters. Indonesian strain of scale carp at sub station Cuttack grew well in plains where it breeds prolifically.

Common carp is polytypic plastic species with a marked tendency to produce varieties and races in response to selective breeding and environmental influences. A large number of varieties with equal strains and races now exist all over the world.

In India there are two strains. The 1st, the Prussian strain which is having Calician ancestor was introduced in 1939 from Ceylon and the Chinese or Eastern strain which was imported from Bangkok in 1957.

Both the strains have now mixed beyond identity.

Characters

1st ray of dorsal and anal modified into spine and both are internally serrated. Two pairs of barbells. Scales all loosely placed. Scales are shining and small in size.

Growth

In tropics, it is cultivated upto altitude of 1000 metre above sea level. Optimum temperature for growth is 20° C to 25° C. Under favourable conditions, it acquire a growth of 2.5 to 5 cms. per month and attains 1 Kg and 2 Kg weight at the end of 1st year and 2nd year, respectively. Stocked at thin rate of stocking higher growth rates are possible.

Under mono culture it gives different growth in different countries depending upon environment. When fed with aquatic or terrestrial grass leaves of cauliflower, cabbage and *Ipomea* at a feeding rate of about 100 gms per day per fish it gives an average weight of 28.5 gms, 25.0 gms, 14.0 gms, and 3.6 gms, respectively in one month.

Food and Feeding

At spawn stages it feeds on nannoplankton, shifting to plankton containing Protozoa and Naupli. From early fry to late fry it feeds on zooplankton-consisting of Rotifers, Naupli, *Cereodaphnia*, *Moina* etc. and phytoplankton containing-*Diatoms*, *Macrocystes*, *Euglena*, *Oscillatoria*, *Volvox*, *Closterum* etc. From fingerling to adult stage it feeds on benthic fauna like chironomids, tubifex and other insects larvae, decayed vegetable matter, detritus of phyto-zooplankton and also other oligochaetes. Adults are bottom feeding, omnivores and readily take to artificial feed. These feed on periphyton and plankton. They collect food by taking bottom mud into the mouth, taking in digestible particulate food and throwing out mud, making water turbid and impair plankton production besides destroying bottom vegetable. It also makes embankments weak by burrowing holes in search of worms and insects.

Maturation

In Russia, it attains maturity between 4 – 5 years. In temperate zones of Europe males ripe at 4 years and females at 5 years and in Israel at 12 months even if the fish has reached a weight of only 100 gms. In tropics of Africa the species mature in 1st year when ½ Kg by weight. In India at Cuttack, Chinese strain species mature at 6 – 8 months age, males maturing earlier.

Fecundity

Fecundity of the species per Kg of fish weight varies from 132,000 to 160,000 eggs in waters of far East Asian countries.

Spawning

In temperate countries in spring once a year, while in tropical waters it spawns throughout year with peak period March to April. Eggs are adhesive and are of two colours. Hatch out in 48 – 96 hours depending on temperature.

Distribution

Autochthonous in freshwater rivers and ponds of China, Russia, transplanted and cultured practically in all parts of the world in temperate and tropical parts including India.

Thriving in China and Russia between 35 degree N to 50 degree N latitudes and 30 degree E to 135 degree E longitudes. It was transplanted to Europe and East Asian countries all over the world.

Economic Importance

Largest cultivable and domesticated fish all over the world. Shows excellent growth in suitable environments and takes artificial food easily. Among carps it is only species which breeds in confined waters forming components of composite culture.

The common carp is highly suitable for seasonal ponds of India. The fish grows to 500 to 750 gms in 150 – 180 days. These ponds get dried in summer and when filled by rain water are used for common carp culture. The bottom of such ponds contain detritus both of plant and animal origin become a source of food to this omnivorous fish.

The common carp is a native of the temperate regions of Asia, especially of China. It has been transplanted into scores of countries, so much so that it now enjoys the status of a virtually global fish and its culture is very widespread.

Common carp bigger than 10 cm thrive on decayed vegetable matter containing bottom dwelling organisms, notably tubificids, molluscs, chironomids. Common carp dig and burrow into pond embankments and sides in search of organic matter. The fish gulp in mud from which digestible matter is sifted and the rest rejected, a habit which often makes pond water turbid.

The growth rate of common carp under culture is a function of the variety cultured, rate of stocking, quality and quantity of feed supplied and competition with other fish in the case of polyculture.

The males have remarkably highly developed testes and some have the abdomen bulging as conspicuously as in the case of gravid females. The weight of testes in some males may be as high as 20 to 30 per cent of the total body weight.

Common carp breeds naturally in its natural habitat, the rivers as well as in ponds and tanks. The eggs are adhesive and the fish requires suitable floating substances for attachment of eggs.

Food and Feeding Habits of Fishes

For successful fish farming, a thorough knowledge of the feeding habits of the selected species of fish is essential to ensure that all available food in the pond is used by the species stocked. As the food resources in ponds are varied, as also the kinds of fish and their feeding habits, a judicious combination of species for rearing would achieve this object.

The majority of our commercial carps are omnivorous and draw their sustenance from more than one source of food in the pond. Among the major carps, catla is primarily a plankton feeder, and is, therefore, confined to the upper column or surface layer, while rohu and mrigal feed at the bottom.

The main items of food are given in the Table 3.1 as shown by an analysis of the gut content of a large number of catla, rohu and mrigal in the various stages of their growth.

Our knowledge about the food and feeding habits of our cultivated fishes is still very limited. All available information is summarized and given on Table 3.2 enable the fish farmer to select the right species of fish which can use the maximum food resources available in his pond.

Several of the cultivated species of carps have very similar feeding habits, the bulk or, at least, an appreciable part of food consumed being sand and mud with decaying vegetable debris. These fihes are non-predators with toothless jaws, and cannot, therefore, bite their food. They can however, swallow food which is crushed with a set of teeth at the throat before it is passed down into the stomach. They have certain structural adaptation suited to their peculiar feeding habits.

Catla has a slightly upturned mouth which enables it to gulp in quantities of water which are filtered through its gills, all the

Table 3.1: Food Consumed by the Major Indian Carps, Catla, Rohu and Mrigal during Various Stages

Name of Fish	*Length in Millimetres*	*Average Percentage of Items of Food Generally Encountered in the Stomach and Gut*					
		Unicellular Algae	*Filamentous Algae*	*Vegetable Debris*	*Animalcules and Water Fleas*	*Insects*	*Mud*
Catla catla (Catla)	11.0 to 20.0	10.0	–	3.0	87.0	–	–
	21.0 to 40.0	9.9	–	2.5	87.6	–	–
	41.0 to 100.0	10.0	–	18.5	70.0	–	1.5
	101.0 and longer	8.3	0.7	38.0	44.1	–	8.9
Labeo rohita (Rohu)	11.0 to 20.0	–	–	20.0	60.0	–	20.2
	21.0 to 40.0	14.9	4.0	7.0	37.1	–	37.0
	41.0 to 100.0	14.0	12.5	41.5	18.0	–	14.0
	101.0 and longer	27.9	0.7	56.0	0.5	0.4	14.5
Cirrhina mrigala (Mrigal)	11.0 to 20.0	19.0	2.0	26.9	33.5	–	18.6
	21.0 to 40.0	22.3	3.0	43.0	15.2	–	16.5
	41.0 to 100.0	25.0	–	55.0	–	–	20.0
	100.0 and longer	26.2	6.7	45.5	–	–	21.6

Table 3.2: Natural Food Consumed by the Important Freshwater Food Fishes of India Mentioned in the Order of Importance

Species of Fish	*Major Items of Food Consumed by*			*Remarks*
	Fry	*Fingerlings*	*Adults*	
Catla catla	Animalcules and Water fleas	Water-fleas, a few plankton algae and some vegetable·debris	Water-fleas, vegetable debris including some algae	Mainly plankton feeder
Labeo rohita	Do	Vegetable debris, microscopic plants, few water-fleas, debris and mud	Vegetable debris, microscopic plants, detritus and mud	Bottom column feeder
Labeo calbasu	Do	Vegetable debris and microscopic plants (mainly), few water-fleas detritus and mud	Vegetable debris, microscopic plants, detritus and mud	Bottom feeder
Labeo fimbratus	Do	Do	Do	Bottom column feeder
Labeo kontius	Do	Filamentous algae and rotting leaves of aquatic plants, unicellular algae, Insect remains, detritus and mud	Rotting leaves of aquatic plants detritus and mud, insect remains, algae	Marginal and bottom feeder
Labeo gonius	Do	Vegetable debris, unicellular algae, detritus and mud	Vegetable matter, microscopic plants, detritus and mud	
Labeo bata	Do	Vegetable debris, detritus and mud, unicellular algae	Vegetable debris, detritus, mud, unicellular algae	Bottom feeder
Labeo sp. ('parel')	Do	Do	Do	
Labeo dussumeiri	Do	No detailed study made	No detailed study made	

Contd...

Table 3.2–Contd...

Species of Fish	*Major Items of Food Consumed by*			*Remarks*
	Fry	*Fingerlings*	*Adults*	
Cirrhina mrigala	Do	Vegetable debris, unicellular alage, detritus and mud	Vegetable debris; unicellular algae, detritus and mud	Bottom feeder
Cirrhina cirrhosa	Do	Vegetable debris, sand or mud and detritus, algae	Detritus, sand and mud; rotting leaves of macrophytes, planktonic algae	Bottom feeder
Cirrhina reba	Do	Vegetable debris, unicellular algae, detritus and mud	Phytoplankton organisms, detritus and mud, rotting leaves	
Osteochilus thomassi	Do	Unicellular algae, detritus and mud	Filamentous algae and debris, detritus and mud	
Thynnichthys sandhkol	Do	Microscopic plants, organic debris and sand	Planktonic algae, detritus and mud	
Barbus carnaticus	Do	Insects, filamentous algae	Filamentous algae, leaves of aquatic macrophytes, detritus, insects	
Barbus dubius	Do	Dipteran larvae, detritus and mud	Dipterna larvae, corixid bugs, microcrustaceans, detritus and mud	Bottom feeder
Barbsu hexagonolepis	Do	Insect larvae, beetles and flies	Aquatic macrophytes marginal grass, gastropod shells	Vegetation and molluscan feeder

Contd...

Table 3.2–Contd...

Species of Fish	Major Items of Food Consumed by			Remarks
	Fry	Fingerlings	Adults	
Cyprinus carpio	Do	Water-fleas, few algae, vegetable matter	Vegetable matter and detritus, unicellular algae, Insects	
Carassius vulgaris	Do	Animalcules and water-fleas	Animalcules and water-fleas, vegetable debris, unicellular algae	Plankton feeder
Tinca tinca	Do	Water-fleas, filamentous algae, vegetable debris	Filamentous algae, vegetable debris algae	
Etroplus suratensis	Do	Animalcules and water-fleas, filamentous algae, unicellular algae	Diatoms, filamentous algae, water-fleas insects, leaves of aquatic plants	
Osphronemus gurami	Do	Insects, crustacean	Macro-vegetation Insects	Vegetable feeder
Lales calcarifer	?	–	Fishes, prawns	Predator
Anabas testudineus	Animalcules and water-fleas	Insect larvae, water-fleas	Insect, water-fleas, vegetable debris fish	Predator
Mugil cephalus	?	–	–	
Mugil sp	?	–	–	
Mugil corsula	Animalcules and water-fleas	Microscopic plants, water-fleas, detritus and mud	Detritus and mud; unicellular and filamentous algae, insects	Predator

Contd...

Table 3.2–Contd...

Species of Fish	*Major Items of Food Consumed by*			*Remarks*
	Fry	*Fingerlings*	*Adults*	
Wallago attu	Water-fleas insects, fish fry	Insects, other fish fry and fingerlings	Small and medium size fish	Predator
Mystus seehala	Do	Do	Do	Predator
Clarias (magu)r	Water-fleas (?)	Insects, ostracods, debris	Prawns, shrimps, insects	Predator
Hateropnuestes fassilis	Do	Insects, ostracods	Insects, ostracods, debris and algae	Predator
Chanos chanos	?	Microscopic and filamentous algae	Microscopic and filamentous algae	Plankton feeder
Megalops cyprinoids	Animalcules and water-fleas	Insects, zooplankton fish fry	Small fish, insects	Predator
Channa marulius	Do and insect larvae	Insect larvae, fish fry	Small and medium size fish	Predator
Channa striatus	Do	Dipteran larvae, zooplankton, fish fry	Small fish like Barbus spp. Chela and Esomus	Predator

particulate organisms which give it sustenance being retained. It seems generally incapable of picking up food from the bottom of the pond or of browsing on weeds.

In rohu the mouth is a little sub-terminal in position with tick fimbriated lips. The structure of the gill rakers shows that they are not adapted to filtering minute plankton organisms. The structure and position of the mouth and lips are adapted to sucking out the debris from the bottom and for browsing on weeds.

In mrigal and bata, the mouth is clearly terminal with thin lips, enabling them to pick up the bottom mud and debris easily. The gill rakers are ill-adapted for filtering water and retaining particulate organisms.

These differences in the structure of the mouth and the intestine appear only in the early fingerling stage. The highly coiled intestine, characteristic of the adult fish, seems to be formed later when the baby fish, with its straight tube like intestine resembling that of adult predatory fishes, changes its feeding habits. There is hardly any difference in the position of the mouth also in the tender fry of most of the cultivated species of carps. That the body fish are exclusively animal plankton feeders in the first few days of their life is proved by the structure and position of their mouth and intestine. As the young fish grows, the intestine becomes more and more coiled, and with the change the position of the mouth also shifts. Feeding habits thus differ at different stages.

Carps which pick up or dig out their food matter from the mud at the pond bottom have the lips somewhat protrusible as in *Barbus dubius*. This fish feeds largely on worms, insects and snails at the bottom. With their protrusible lips they scoop out the bottom silt in search of prey, making small shallow pits in the process. If such small pits are found in shallow rivers margins, it is clear indication that *B. dubius* has been feeding in that area. Some of the minor carps like *Rohtee* spp. (catio and vigorsi) and Pearl Spot (*Etroplus suratensis*) have similar feeding habits. The latter, though not a marked predator, bites off portions of aquatic weeds and algae with the teeth in its mouth. *Gourami* has also similar habits.

In predatory fishes, the mouth is generally large and provided with a number of teeth, which, however, are not so much for chewing the prey as for preventing its easy escape. With its terminal mouth, the predator can easily snap at and catch relatively larger victims. Unlike the non-predators, the intestine is very short with a distinct bag-like stomach. The Large Goby (*Glossogobius giuris*), the Murrel (*Channa* spp.) and the freshwater shark (*Wallago attu*) are familiar instances.

Chapter 4

Culture of Indian Major Carps and Exotic Carps

Indigenous Culture Practices of Carps

The annual seasonal collection of millions of early fry of Indian major carps (IMC) from rivers in the States of West Bengal, Bihar, Uttar Prasad and Orissa in India and their stocking in nursery ponds for rearing, is a very important step in the fish cultural practices. Different types of ponds are utilized as nurseries and different methods of 'preparation' of the ponds are followed. There is, however, very little reliable information about these practices, most of which constitute family secrets in peasant families handed down from generation to another generation. It is a common practice to purchase and release some fry of fingerlings in ponds without any attention being given on the fate of the stock. Only a few attempt to clear the pond of unwanted fishes of predators, while still fewer people manure the pond or feed the fish artificially. Reliable data on such important aspects of fish culture as the density of stocking for given waters and the percentage of survival of the different species at different stages, are lacking. Even an expert fish farmer is unable to ensure a satisfactory yield of fish in his pond. Reports of total failures of nurseries managed even by experienced fish farmers are not uncommon.

Against this background, the various fish culture techniques outlined and suggestions made, which are based on experimental data and repeated field observations, should be of particular help to the fish farmer and the extension worker.

Every area of perennial natural waters has a crop of fish in it. But every type of water whether perennial or seasonal (long or short), can be utilized for raising a better crop of fish by adopting proper methods of cultivation.

Ponds vary in location, size, depth, soil, shade, etc., besides in the qualities of water, fish food resources and natural stocks of fish. In other words, no two ponds can be considered exactly alike for fish production, because each has its own physical and biological characteristics. Though most ponds fall into well known categories in respect of their fish production or physico-chemical conditions, they require special treatment for increased fish production. Different categories of ponds are required for cultivating different species of fish and also for rearing the different stages of life of each species. The important phases of carp culture and the types of ponds required for each phase are given in Figures 4.1–4.3.

Hatchling Pits

These are to be located close to the breeding bundhs (as in Midnapore and Bankura districts of West Bengal) from where fertilized eggs are collected in large numbers. The size of the pits may vary, but 2.5 m × 1.5 m × 1.0 m is a convenient size.

Phase of Carp Culture	*Category of Ponds Required*
Hatching of laid eggs collected from rivers or bundh-type tanks	Hatching pits, 2.5 m × 1.5 m × 1.0 m, with rectangular, coarse and fine-meshed miniature cloth tanks with or without arrangements for running water.
Nursing the hatchlings during the first two weeks to one months	Nursery ponds, preferably shallow (1 to 1.5 feet deep) small, seasonal; in perennial, 1.5 to 2.0 m deep, long and narrow to facilitate netting
Nursing advanced fry or early fingerlings to the late fingerling size (10 to 15 cm)	Rearing ponds, long-seasonal or perennial, depth 1.5 to 2.0 m, preferable long and narrow.
Rearing fingerlings to adult size	Stocking ponds, perennial, 1.75 to 3 m or even deeper, preferably narrow (up to 30.5 to 45.75 m) and long.
Inducing mature carps to breed	Breeding ponds-bundh type or ordinary and small, with weeds for species which breed in ponds.

A series of such pits, preferably interconnected so as to allow the water to flow from one pit to the other, should be available. There should preferably be a continuous and slow flow of water either from a natural source or from an artificially created flow (by a pump) in and out of the pit, so that the eggs when laid for hatching are well aerated.

In each pit, a smaller rectangular cloth tank, 2.0 m × 1.0 m × 0.5 m, made of cheap coarse cloth should be tied to bamboo poles fixed in the pit. A similar open tank, but of smaller dimensions (1.75 m × 0.75 m × 0.30 m) made of round-meshed mosquito curtain cloth may be fixed within the outer course cloth tank, with its upper edges projecting out of the water surface by 7.5 to 10 cm only.

Smaller hatching pits, without the cloth tanks or hapa inside, are commonly used near the bundh-type tanks as in Midnapore district. The survival of hatchlings in these pits is reported to be low as usually several hatchlings get injured while being collected for such pits.

Nursery Ponds

Wherever, possible, the nursery ponds should be located close to the collecting ground so that fresh fry could be released directly into them without conditioning and transport. Small ponds, 15.25 m × 15.25 m, which can be easily controlled. They should be seasonal and shallow, with not more than 0.915 to 1.22 m of water, and should dry up completely during the summer months. Even short-seasonal ponds containing a depth of 0.915 to 1.22 m of water for about three months of the monsoon season may perhaps be better for this purpose, as they have the additional advantage of all the fish enemies being eradicated when this water dries up.

It helps to eradicate the fish predators, but also to improve the general hygiene of the pond bottom by getting rid of the poisonous gases, if any, and enriching the same by contact with air. If the ponds are not seasonal, they should preferabiy have outlets for dewatering, so that they could be drained when required. In the absence of such facilities, the pond should be dewatered by pumping out water once every four or five years.

They should, as far as possible, be located away from the river banks so as to avoid fluctuations of the water level in the river during flood and dry season.

Rearing Ponds

They may be perennial or seasonal, retaining water for a long period. The advanced fry or early fingerlings should be kept in such ponds for two to three months when they attain advanced fingerlings size (10.16 to 15.0 cm in length). Netting greatly facilitate if the ponds are long and narrow, and have gently sloping sides and the depth of water does not exceed 1.25 m. Old silted ponds are not suitable for rearing fry. Paddy fields holding water to a depth of 0.46 to 0.61 m for two to three months may also be utilized for rearing early fingerlings.

Stocking Ponds

All perennial ponds, a third of an acre or larger in area, and over 1.8 m in depth, are better used as stocking ponds. As all available standing waters are to be used for fish culture, the size of a pond need not be taken into consideration, but if new ponds are constructed, a long run for the fish and easy netting facilities may be borne in mind. Long and narrow ponds, not exceeding two to three acres in extent, may prove to be of a manageable size for netting by a group of six fishermen using a relatively small net of 30.5 m or 45.95 m in length.

Breeding Ponds

The most important of these are the bundh-type tanks in which the Indian major carps (IMC) are known to breed. These are minor irrigation tanks, some perennial and others seasonal. The rain water, flowing down the sloping catchment area, is suitably blocked in a natural or excavated depression by creating an earthen bundh across the outflow. With the continued accumulation of water in the depression, the level of water rises, submerging the adjoining shallow areas or fields. A weir is constructed for the overflow of water when the depression is full, and the adjoining extensive shallow fields are submerged.

Breeders of carp are reared or introduced from else-where into the pond. Early in the monsoon, the pond gets rapidly filled up, and later the rains cause sudden floods by enhanced flow of water from the catchment area. Whether the water flows out through the outlet or not, the fish move up to the shallow, inundated grounds or fields (30 cm or 60 cm in depth), for the purpose of spawning.

Cultural Methods

Seed Collection, Handling and Mortality

Our major rivers are the most important sources of seed of Indian carps. From these, millions of eggs, hatchlings and tender fry are collected every year during the monsoon months. A fair quantity of eggs is also collected from the bundh-type ponds of West Bengal by the simple process of disturbing the ground on which the eggs are laid so as to float them up and catch them with simple cloth nets.

Hatchlings and fry are collected mainly from rivers in specially made fry nets. These are conical or funnel-shaped nets made of cheap coarse cloth so as to allow easy filtering of water.

The length of the net varies (5.79 to 6.71 m) at different places. It is open at both ends. The narrow tapering end is fitted with a ring (22.86 to 30.48 cm in diameter) made to reeds or cane. The mouth of the net is provided with tape or rope for proper fixation on to the poles. The capacity of the net is often enhanced by adding two "wings" at the mouth.

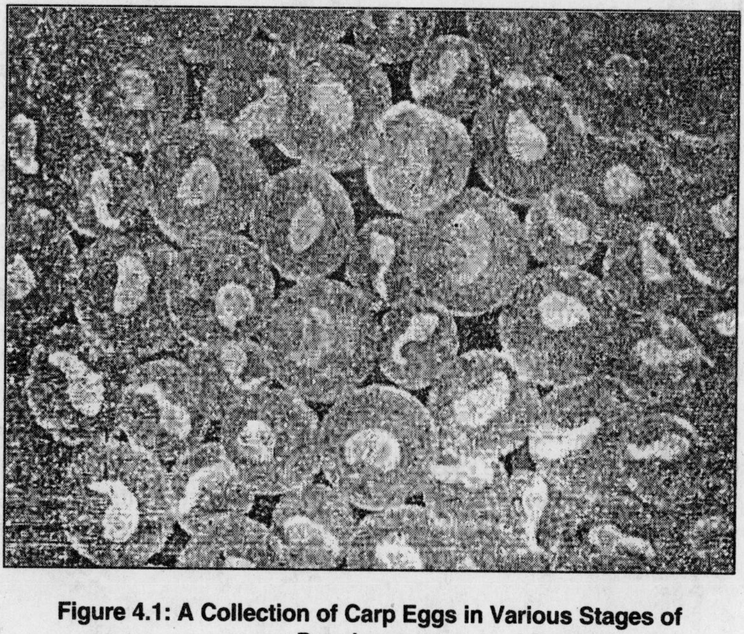

Figure 4.1: A Collection of Carp Eggs in Various Stages of Development

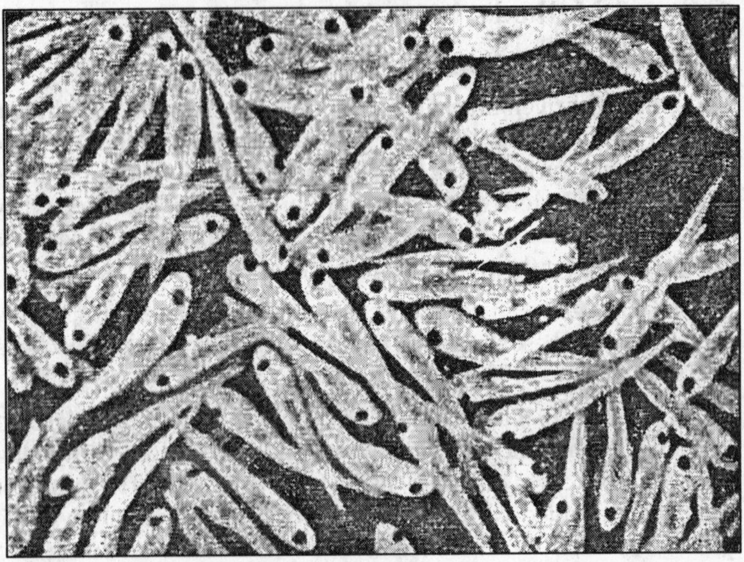

Figure 4.2: A Collection of Carp Fry at the Stage at which they are Ordinarily Collected from Rivers

The net is fixed in shallow water where the flow is slow or gentle. It is fixed to poles with the mouth facing the current. The hind portion will drift in the direction of the current in which position it is fixed to the poles, just below the surface. The upper edge of the mouth of the net will be seen above the water. To the hind ring of the net is tied a tailpiece cloth, often two feet long and a foot deep, and tapering posteriorly. This tailpiece, usually called *gamcha*, is also tied with the upper edge just above the water. The carp fry drifting with the current pass through the mouth of the net, get collected in the *gamcha* and are periodically removed and temporarily stored in small pits or in miniature cloth tanks fixed in the river itself.

A critical study of this common practice will reveal that during collection the tender fry get injured in several ways. If the net is not properly fixed, resulting in resistance to the fast current, several fry are either injured in the net or die. The damage to the fry can be greatly prevented if the net is fixed mouth first, allowing the tail

Figure 4.3: A Developing Carp Egg with the Embryo just Differentiated; same with Advanced Embryo Inside

portion to drift in the direction of the current and then fixed to the poles in the fully stretched position.

Usually, flood waters wash down large quantities of debris which are often caught in the net. These accumulate in the *gamcha*, and physically choke and injure the tender fry. This choking can be appreciably minimized if the form of the posteriorly tapering *gamcha* is changed to an open rectangular tank-like one, roughly 0.75 m × 0.3 m × 0.45 m and by introducing a net screen of about half inch mesh at the mouth of the net. The large-sized debris will be caught against the net and can be removed from time to time so as to allow the free flow of water and the movement of the fry contained in it.

The collection of fry from the net is usually cleared at long and irregular intervals, more particularly at night. When the collection is heavy and the fry are not cleared at regular, shorter intervals, there is bound to be considerable overcrowding in the *gamcha*, particularly when the choking due to accumulation of debris is not cleared. It is, therefore, essential to have the nets cleared at regular fixed intervals.

In collecting the fry from the *gamcha*, even trained fishermen lift it very often out of the water to drain and transfer the whole mass of fry and debris together to a bucket of water. This results in injury to the delicate fry from which recovery is often impossible. The correct method of clearing the *gamcha* is to lift up its nearer end by passing a hand underneath, driving the fry to the farther end, and passing the hand forward till all the fry congregate in a limited volume of water which can then be scooped up with a small vessel to be transferred to the water in the bucket.

The fry thus collected are carefully separated from the debris by passing the collection through a fine-meshed sieve through which the fry will pass. When the sieve is placed on the mouth of a hundi and the collection poured over it, the debris is caught on the sieve allowing the fry a fall into the hundi, a process entails considerable injury to the fry if the sieve is not kept immersed in water.

If the fry are to be temporarily stored in hundies, pits, or hapas, when they are collected, care should be taken not to overcrowd them with the fry. Overcrowding results in quick loss of dissolved oxygen and accumulation of carbon dioxide in the water and finally in the death of the fry. Even if the exact number of fry that could be safely kept in any particular container is not known, a careful watch maintained over the fry would reveal gradual overcrowding of the fry on the surface, which is an indication that the water needs partial or complete renewal.

This is easily done by pushing a coarse-meshed cloth down into the centre of the water in the hundi and the quickly removing the water with a cup without handling the fry. Fresh water is forthwith sprayed or gently poured into the hundi to replace the foul water removed already. In spite of all these precautions, some weak fry may succumb to the strain during collection.

The hatchlings and fry have often to be stored for a few days, until they are ready for transport in hundies to distant places. Both in storage and transport, they are liable to suffer injury or adverse conditions, ultimately leading to death.

When the conditions of water at the place of collection, in the conditioning pit, or in the container are significantly different and the fry are suddenly transferred from the one the other, they are likely to be adversely affected. The safest course in such cases would be to gradually mix the two waters before transferring the fry.

On bright sunny days, the water in the shallow conditioning pits or containers is likely to get heated up, particularly when they are not in shade, resulting in large-scale mortality of fry. The containers should, therefore, be kept either in the shade of a tree or protected by a temporary shed erected for the purposes over the hapa pits. Overcrowding in the hapa or the container should be avoided to prevent depletion of oxygen which results in mortality.

In hapas or containers with stagnant water, the weak and injured fry are susceptible to early attacks by fungi or bacteria. Whenever possible, running water should be readily available in hapas to prevent this. If such an arrangement is not feasible, common salt may be added to the water to any way adversely affect the fry but the fungal and bacterial infection will be checked and better hygiene will prevail in the hapa.

Conditioning of the fry fingerlings is a common practice in many parts of India. This consists of keeping a number of them in a limited volume of water for a definite period during which the gut contents are voided through the vent and the fish get accustomed to the limited volume of water in the container.

As conditioning involves starvation of both the fry and fingerlings, it should not be prolonged so as to prevent mortality.

The larger fry jump against the net or container in attempting to escape and injure themselves, thereby rendering themselves liable to fungal and bacterial infection.

Although sieving the fry collection results in the separation of most of the larger fry of predatory fish from the carp, the fry of smaller predatory species and some insects larvae escape through the meshes of the sieve into the container to feed on the carp fry. There is little doubt that in the unsieved collection of fish fry the carp fry suffer injury and death due to the presence of a large number of predatory fish fry and insect larvae in the limited space of the hapa of hundi. The fry are liable to suffer injury if netting them out and measuring are not done with the utmost care.

Seed Transport and Mortality

The fry the fingerlings are transported in earthen hundies or in tin carriers of various capacities and shapes and sometimes in sealed containers with oxygen. At the temperature that prevails during the rainy season, the dissolved oxygen requirements of fry and fingerlings carried in a 20 liters container with river or pond water are not unduly high. The number of fry of fingerlings that could thus be safely transported.

This data is in respect of a closed container at rest. When the fish seed is transported in open containers, the constant jolting brings about an appreciable replenishment of atmospheric oxygen in the water. This will afford a longer safe period for the fry than is shown in the table.

Table 4.1

Length of Fry	*No. per Hundi*	*Duration of Transport (in hours)*	*Percentage of Mortality*
1.25 to 2.0 cm	1,500	Up to 24	2.0 to 5.0
"	1,200	Up to 36	2.0 to 5.0
2.0 to 2.5 cm	1,000	Up to 20	2.0 to 5.0
"	800	Up to 30	2.0 to 5.0
2.5 to 5.0 cm	500 to 800	Up to 24	10.0
2 to 3 inches	200	Up to 8	10.0

Table 4.2

Volume of Water in the Container	Initial Dissolved Oxygen Content	Size of Fry to be Transported	No. of Fry to be put in Each Container	Approximate Safe Period in Minutes during which Transport can be Effected
20 liters	4 ppm	6-7 mm	50,000	19
"	"	"	30,000	31
"	"	"	20,000	47
"	"	15 to 20	1,000	40
"	"	"	500	80
"	"	30 mm	300	120
"	"	"	150	240
"	5 ppm	6 to 7 mm	50,000	25
"	"	"	30,000	42
"	"	"	20,000	62
"	"	15 to 20 mm	1,000	60
"	"	"	500	120
"	"	30 mm	300	165
"	"	"	150	330
"	6 ppm	6 to 7 mm	50,000	31
"	"	"	30,000	51
"	"	"	20,000	77
"	"	15 to 20 mm	1,000	75
"	"	"	500	150
"	"	30 mm	300	207
"	"	"	150	414

In rail transport, using ordinary open earthenware hundies with about 22 liters of water, part of which is renewed repeatedly enroute, the number of fry that can be transported safely.

When the dissolved oxygen content in the water goes below the minimum required, the fry show signs of distress and come up gasping to the surface. This is an indication that the water should be renewed immediately. This is done by carefully removing some water from the container and replacing gently with an equal volume of fresh untreated water. Chlorinated water should not be used, as it is harmful to fish.

As fry or fingerlings which die in transit begin to rot and foul the water, making it unsuitable for the other live fish, the dead fish should be removed periodically. Likewise, specimens infected with fungi or bacteria and which swim feebly with partly damaged whitish fins, are likely to infect the healthier fry in the container, they should also be removed.

When fry are transported over long distance in open vehicles, the metal containers and the water in them get heated up. This results in the death of the fry. If cheap woolen or gunny coverings are provided for the containers and kept moist by sprinkling water on them, the temperature of the water can be kept down. Alternatively, earthen pots which keep the water cool may be used as containers. As the transport of fry of different sizes in the same container also contributes to the mortality of fry, they should as far as possible be sorted out according to size in different containers. The fry of predatory fish and insects and their larvae should be scrupulously eliminated from the containers.

The fry can successfully be transported by air in sealed containers with oxygen for long distance transport. New, 15 liters kerosene tins fitted up with air-tight screw cap lids for filing in water and fish, and with tubes for letting in oxygen from an oxygen cylinder and displacing water, have been successfully used. When the tin is filled two-thirds with water and a third with oxygen, the number of fry that can be safely transported is given below.

Length of Fry	*No. per Hundi*	*Duration of Transport (in hours)*	*Percentage of Mortality*
1.25 to 2.0 cm	800 to 1,100	12 hours and 8 hours by road	5 to 12
4.0 to 5.0 cm	400	24 hours	30
2.5 cm to 3.0 cm	325	16 hours	3

The fry and fingerlings are to be conditioned for about six hours before packing in the tins. Fingerlings over two inches in length are unsuitable for economical transport over long distances.

Seed Stocking, Rearing and Mortality

If there are appreciable difference in the temperature and other conditions between the water in the container and that in the nursery pond, the fry released in the latter are likely not be directly released into the pond. The container with the fry should be kept partly immersed in the pond for some time allowing the temperature to equalize. Small quantities of the pond water should be occasionally sprinkled into the container. After about half an hour of this acclimatization the container may be tilted so as to allow the fry to escape into the pond.

The fry stocked in the pond should be provided with enough quantities of suitable food. The general practice of releasing the tender fry into the pond without previous "preparation" to receive the stock followed in many places is definitely harmful and uneconomical. A systematic preparation of the nurseries is essential for ensuring satisfactory survival of the seed.

The fry of predatory fish, which take a heavy tol of carp fry, are often inadvertently introduced into the pond along with the latter. Improved methods of collection and careful segregation of all unwanted fry by careful sieving alone will help reduce mortality of valuable fish seed.

Apart from the predatory activities of the numerous aquatic insects, minnows and other unwanted fishes in the water, the cannibalistic tendency of the carp fry themselves in their early stages has also to be remembered when fry of different sizes are stocked together in the same pond at intervals of two to three days, as some of the smaller and weaker fry are eaten by the larger and stronger ones. To avoid this, fry of the same size should, as far as possible, be used for stocking in a single day. Further stocking in the same pond should, as a rule, be avoided at least till 10 days, of the first stocking. A second stocking, if found necessary, is best done after the first stock has been removed after a lapse of about 12 to 14 days.

In certain ponds, the microscopic floating algae, multiply rapidly and turn the water bottle-green in colour. Such blooms occurring on sunny days utilize the carbon dioxide in water, releasing oxygen quickly to the pond of supersaturation. Oxygen, though essential for life, is harmful when present in excessive quantities. Tender carp fry present in such a water are soon affected. The excess oxygen usually accumulates in the form of small bubbles in the gut and other body spaces, including blood vessels, making the fry unstable and float on the surface. The continued accumulation of gas eventually results in the bursting of the stomach or gut, and death. As already stated, algae are not preferred as food by the tender fry. The release of fry into ponds having thick algal blooms is, therefore, not advisable except when the bloom is on the decline. Advanced fingerlings may, however, do well in such ponds.

Overgrowth of submerged or floating weeds and filamentous algae are as unsuitable as algal blooms for carp fry in nursery ponds, as they are not used as food. In a pond full of such weeds, the plankton contents on which carp fry generally thrive will also be very poor.

Thus several favourable factors should prevail in the water from the movement carp fry are released in the ponds up to a period of week or 10 days thereafter, so that a satisfactory growth of the fry may be obtained. Unless suitable steps are taken to create and maintain such favourable conditions during the critical period, the majority of fry will perish, making fish culture an unprofitable concern.

Predatory Fishes in Nursery Ponds

The common predatory fishes met with in nursery and stocking ponds have different local names. They are *Channa marulius*: sal (West Bengal), bhor (Bihar); *Channa striatus* : shol (West Bengal), Assam; sowra (Bihar); sowarah (Andhra); *Channa stewarti* : dudu-cheng or tel cheng (West Bengal); *Channa punctatus* : taki, lata (West Bengal); *Channa gachua* : cheng (West Bengal), (Assam); chainga (Bihar), *Anabas testudineus* : koi (West Bengal), (Assam); kobhai (Bihar); *Nadus nandus* : nandus or dudhurkal (West Bengal); bao-vaadhi (Assam); vaadhul (Bihar), (Uttar Pradesh); meussoassal (Punjab); bobari gossipora (Orissa); septi (Andhra); andi kalla (kerala); *Glossogobius giuris* : beley bhalia (West Bengla), bhula (Bihar); *Notopterus notopterus* : pholui (West Bengali); golhi (Bihar); pholi (Uttar Pradesh); *N. Chitala* : chital (West Bengal); mohi (Bihar), cittala (Orissa); *Lates calcarifer* : bhetki (West Bengal); fitada, bekti (Orissa), *Heteropnuestes fassilis* : singhe, (West Bengal), singhee, sheenee (Assam); sinhi (Bihar), singi (Uttar Pradesh); *Clarias batrachus* : magur (West Bengal); mah-gur (Assam); mangri (Bihar); *Wallago attu* : boal

(West Bengal, Assam), silond (Punjab); *Pangasius pangasius* : pangas (West Bengal, Assam), jellum (Orissa); *Ompok* spp. : padda (West Bengal); chechora (Bihar); *Mystus seenghala* : (West Bengal, Assam, Bihar); pogal (Uttar Pradesh); tengra (Punjab) aadi (Orissa), shengal (Andhra); *M. aor*; air (West Bengal); seengala (Punjab); aadi (Orissa); *M. cavasius* : kavasi tengra (West Bengal); *Anguilla benghalensis* : kuchia (West Bengal); *Amphipnous cuchia* : kuchia (West Bengal).

Predatory fishes, are almost invariably present in ponds, though in varying number. Even in ponds completely isolated from other sources of water, some of the predatory fishes like the murrels, *Anabas*, and others, crawl in over wet ground from neighbouring waters. And once they get into the pond it is extremely difficult to eradicate them by the usual netting operations. All the predators feed on carps, adult specimens preying upon fingerlings and yearlings, and the young ones destroying the carp fry. A fry of *Wallago attu* kept in an aquarium with 100 carp fry, with the number devoured by it made good every 24 hours, has been reported to have consumed 1,096 fingerlings of carps in 40 days.

Some of the predatory fishes like the murrels, *Anabas, Nandus, Heteropnuestes, Clarias* and *Notopterus* breed in ponds a little prior to or at the commencement of the monsson. Their young ones, feeding on the plankton in the nursery ponds, grow so fast that when the carp fry are ready to be introduced into the pond, they are large enough to destroy them in large numbers.

Young fry of Chital (*N. chitala*), Pangas (*Pangasius pangasius*), etc., which are very often caught in the rivers along with the carp fry have been found to be highly destructive. In most of them the stomach is gorged with carp fry which very often from 100 per cent of the feed. Even the spiny eel (*Mastacembelus pancalus*) with its small mouth is partial to carp fry. The young ones of Chital, the larger riverine catfishes like *Wallago, Silonia* and *Mystus*, the murrels and the climbing perch (*Anabas*) are exceedingly destructive to tender carp fry.

Other Predators in Nursery Ponds

Aquatic insects usually abound in all fresh waters. Some of them, particularly the back-swimmers (Notonectids), occur in thousands during the rainy months in the manured nursery ponds

**Figure 4.4: Enemies of Carp Fry–
The Common Predatory Aquatic Insects**

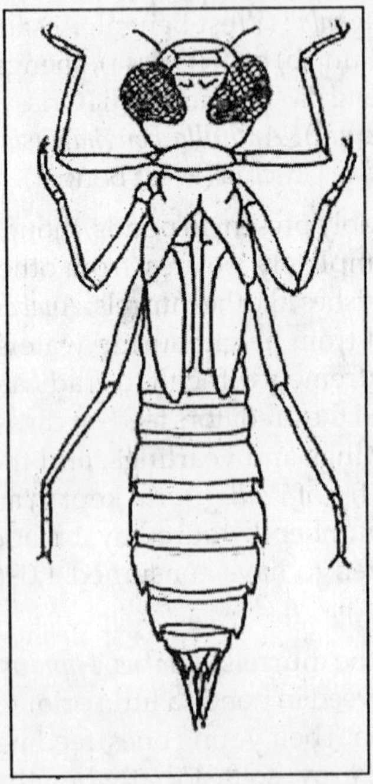

The Dragon-Fly Nymph

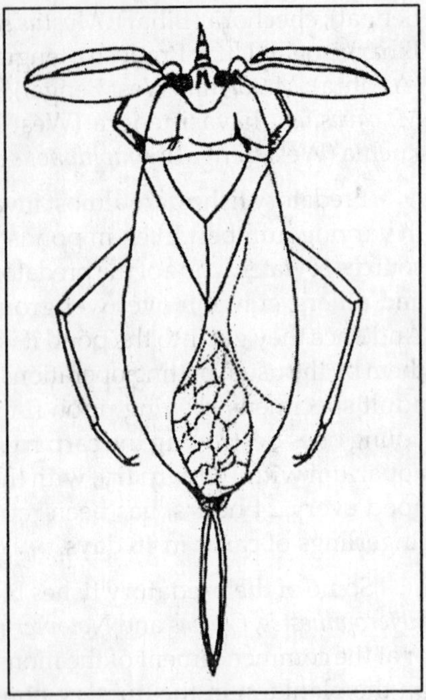

***Nepa* sp.**

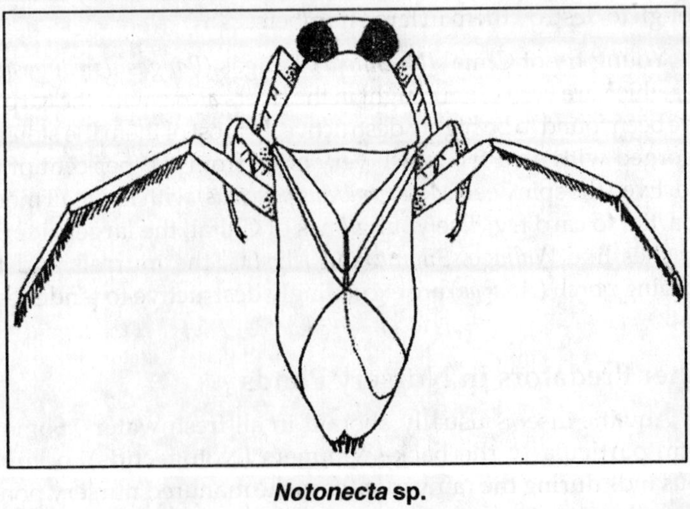

***Notonecta* sp.**

Contd...

Figure 4.4–Contd...

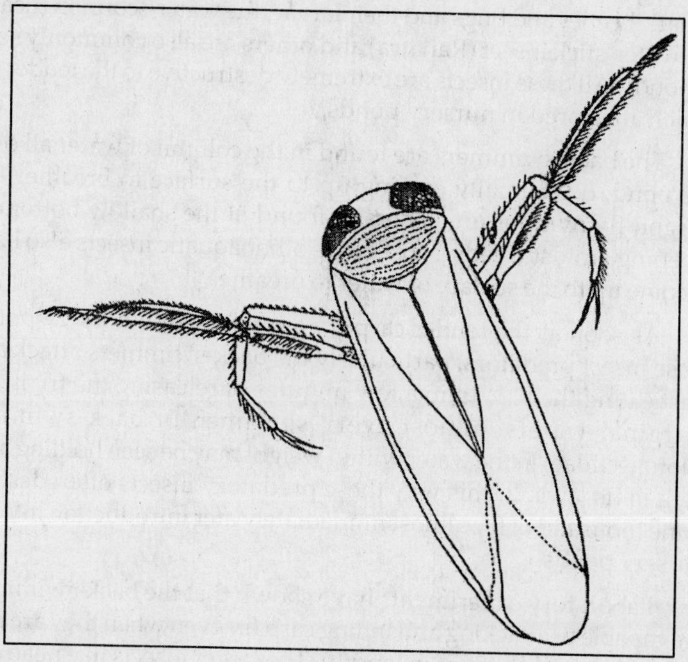

***Corixa* sp.**

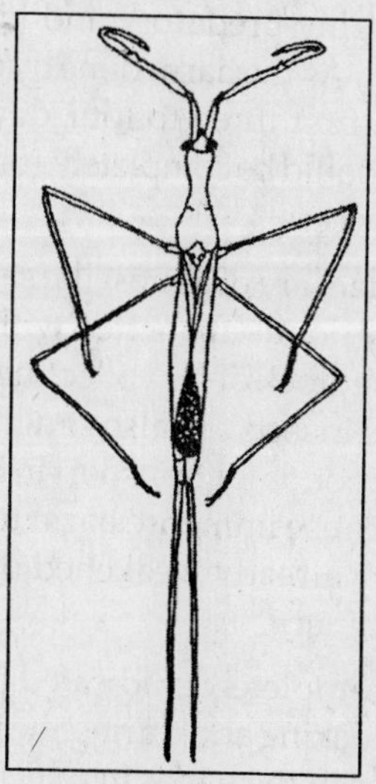

***Ranatra* sp.**

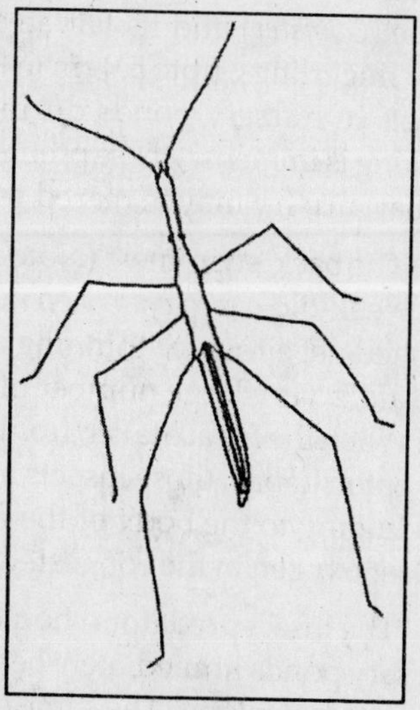

***Hydrometra* sp.**

in Bengal, Bihar and Orissa. Several species of dragon-fly nymphs, aquatic beetles and bugs and their larvae, the water-scorpion (neap), the water-stick insect (Ranatra) and others are also commonly seen in ponds. All these insects are extremely destructive to the tender fry which are stored in nursery ponds.

The back-swimmers are found in the column of water all over the pond, occasionally coming up to the surface to breathe. The dargon-fly nymphs are generally found at the shallow bottom or creeping amidst weeds. Most of the other aquatic insects also have to come up to the surface of water to breathe.

As soon as the tender carp fry are released in nursery ponds these insect predators, particularly the back-swimmers attack and start eating them. Within a few minutes of releasing the fry in the marginal waters, almost every specimen of back-swimmer (Notonectidae) in the water within weight may be seen holding a fry each in its jaws. In this way these predatory insects often destroy some thousand of fry almost immediately afer introduction into the nursery ponds.

Laboratory experiments have shown that the back-swimmers are capable of attacking and killing carp fry even when they are four to five days old (1.25 cm in length). The larger insects like neap and Ranatra can kill larger fry even up to an inch in length. The olive beetle (*Cybister*) and its larvae are highly predatory and kill even early fingerlings, upto 5.0 cm in length. As the harm done by aquatic insects in nursery ponds during the first three to four days after stocking is immense, these predators should be completely controlled so that carp fry maybe reared with success.

The back-swimmers (Notonectidae) are of almost the same size as a week-old carp fry. When nursery ponds are netted with a fine-meshed net either for thinning out the stock of for collecting the fry for sale, a very large number of these insects are also caught in the net. When the fry are taken from such ponds for transport in hundies or kept in hapas, those insects again cause immense harm to the fry by clinging to the body of the fish fry already weakened owing to lack of oxygen in the congested container.

The insect predators should be completely eradicated from the nursery ponds immediately before stocking and during the first few days after stocking. The same applies to the hapas in which the fry are temporarily stored, and to the hundies in which they are transported.

Weed Fishes in Ponds

As no attention at present in being paid to the careful preparation of the pond in carp culture, the nursery rearing and stocking ponds generally contain a varying population of minor carps, carp minnows and predatory fishes. The several species of small fishes belonging to different families are collectively called "weed fishes" as they like the weeds in the field, are harmful to cultivated carps. Multiplying rapidly in the ponds even before the commencement of the rains, their young ones get sufficiently grown, not only to compete with the major carp fry for food but also to directly prey upon them.

Laboratory experiments have clearly shown that all species of weed fishes, big or small take a direct toll of the tender carp fry and devour them as they do with water-fleas when introduced into water. Field experiments on the rearing of carp fry in ponds containing weed fishes have also clearly demonstrated that only a small percentage of the fry introduced actually survives to reach the fingerling stage.

The early fry of the Indian major carps are to some extent cannibalistic in habit. The weaker ones are easily preyed upon by the stronger and larger ones. This habit is more pronounced when the fry are conditioned in hapas or hundies for some considerable time, as in the absence of adequate food they feed on the dead or dying ones.

Eradication of Predatory and Weed Fishes

Among the predatory fishes, the murrels, *Anabas*, *Heteropnuestes*, *Clarias*, etc., are hardy species, breathing atmospheric air but capable of staying at the bottom of the pond or buried in mud for considerable periods. It is, therefore, extremely difficult, if not impossible, to eradicate them from the ponds by netting. As a rule, the predators are fished out by stirring up the bottom silt, when the water level in the pond is very low, and then quickly drag-netting the pond. This repeated at intervals of several days. Large numbers of predatory fishes as also the majority of carps and other fishes can be caught in this way. As a complete eradiation of the predators is seldom possible by the method, those that have escaped the drag-net soon multiply in the pond.

In deeper waters, netting operations are hardly effective. The murrels, for instance, burrow in the bottom mud and escape the nets. Complete dewatering in summer, partial desilting and exposure of the pond bottom to the sun for some days are the only effecting methods for eradicating the murrels from shallow ponds.

In deeper ponds which can be dewatered only at great depths, a thorough and repeated netting should be followed by systematic baiting with hooks. Carp minnows (*Chela, Amblypharyngodon, Barbus,* etc.) serve as very convenient live baits. Short, baited hoots suspended from floats scattered all over the ponds, or a long series of hooks suspended from a long floating line at fixed intervals, will serve the purpose. In relatively small ponds, the two ends of the long line are fixed on opposite banks, but in larger sheets of water they are tied to float provided at either end, or with one end fixed and the other floating. Baited and set at dusk, the line is left overnight and drawn up the next morning. Repeated use of such baited hooks can effectively check the predatory fish population in the ponds.

As stated above, some of the predatory fishes (murrels and the Climbing Perch) breed in ponds, with a peak period in April-May. The gregarious young of murrel move about in a single large shoal, often guarded by the parent fishes. If a hook baited with a live minnows or worm is introduced near the shoal, the parents will forthwith take it. As soon as the parent fish are removed in this manner the fry could also be carefully netted out with a piece of cloth. This method can be effective only during the breeding months when the broods can be located by their conspicuous reddish colour.

Control of Insects in Ponds

An effective control of aquatic insects in nursery ponds is one of the fundamental and essential steps to ensure successful rearing of carp fry in them. In the absence of suitable control measures, the rate of survival of fry will largely depend on the population of insects present in the pond at the time of stocking.

A simple method of control of insects is to drag-net the pond repeatedly with a fine mesh cloth net a day or two before the carp fry are to be stocked in them, and destroyed all the insects thus caught. This substantially increases the survival of fry and production of fish in the pond. This operation should be carried out very carefully so that the water is not unduly disturbed and the favourable ecological conditions for the zooplankton population in the water are not upset. As the insects cannot be eradicated completely by this operation, the aim will be to reduce their number to an appreciable extent.

Many aquatic insects (beetles and bugs) can migrate from one pond to another pond as they can fly out of water. Even if a pond is effectively cleared of its insects population, it will be repopulated very shortly from other adjoining waters. The netting out of Insects should be so timed as to leave little interval between such removal and the stocking of the pond with fish fry and the plankton in the pond should be as little disturbed as possible. In view of the above, lime which are lethal to the microscopic animals and render the water toxic for the succeeding few days, cannot be applied for the control of insects in nursery ponds.

Many species of aquatic insects, including the destructive back-swimmers (Notonectids), can be effectively controlled by spraying common vegetable oils. Oils of mustard, coconut, til, linseed, castor and groundnut can be used without any adverse effect on either zooplankton or on carp fry. Linseed oil at 9.18 kg per ha, however, though very effective against notonectides, has a slightly unfavourable effect on zooplankton and also carp fry. It should not, therefore, be applied in nurseries immediately before or immediately after stocking them with fry.

Owing to high viscosity, these oils do not readily spread on the water surface. This can, however, be overcome if any of the relatively less viscous oils like mustard oil is made into an emulsion with about a third of its weight of any cheap washing soap. When sprayed, this emulsion spreads quickly, forming a thin film over the water surface. The back-swimmers, which have the habit of coming up to the surface of the water to breathe air, come into contact with the oil film which chokes the respiratory tubes and brings about their death.

The required quantity of soap is dissolved in a limited volume of water by boiling, and in this solution, when lukewarm, weighed quantity of the oil is thoroughly stirred to form a yellow emulsion. Before the emulsion is sprayed over the water surface, it is diluted with the pond water. Small quantities of the emulsion taken in a cup or can are sprinkled with force over the water surface so as to spread the same over the water.

As the wind will bank the film of emulsion to one side of the pond, many insects may escape the effect. It is, therefore, advisable to choose a still and windless day for spraying. On a windless day, the back-swimmers in a pond can be completely killed within four hours of the application of the emulsion.

The quantity of oil to be applied is calculated on the basis of the water surface of the pond. An emulsion made 9.18 kg of mustard oil and 2.8 to 3.3 kg of soap per ha. of water surface yields satisfactory results under field conditions, but the effect will last only for 8 of 12 hours after which the pond could be repopulated by insects. A nursery pond is therefore treated with the emulsion a day before it is stocked with carp fry. Ponds so treated have given very satisfactory results, the survival of fry being as high as 90 per cent of above. The additional expenditure involved is fully compensated for by the substantial final results in the yield of fingerlings.

As the oil emulsion has no adverse effect on carp fry or plankton and as its effect on insects lasts only for a few hours after application, a supplementary treatment, a day or two after stocking, may be found necessary and even advantageous.

Larger insects like the water-scorpion, the water-boatman and the olive beetle, the larvae of other beetles and bugs and dragon-fly nymphs, are not generally killed by the dose which is fatal to back-swimmers (Notonectids). Laboratory experiments with stronger doses of the emulsion have proved their efficacy in the control of the population of the larger insects also.

Depending on their cost and availability, mustard, coconut or til oil may be used with effect. Castor oil is too thick to make a satisfactory emulsion.

As aquatic insects attack even advanced fry and early fingerlings during netting conditioning and transport, the application of mustard oil-soap emulsion is recommended for ponds containing those phases also.

When a week to 10 day old fry and netted from nursery ponds, they will be mixed with a large number of Notonectids. These should be completely segregated from the fry, not by the casual method of producing a mild artificial current and allowing the fry to swim against it, but by the entire haul being kept in a medium-sized cloth hapa into which a little mustard oil soap emulsion is sprayed on the surface to kill the Notonectids and removing them. If this precaution of separating the Notonectids from the fry is not taken before the latter are introduced into hundis or tin carriers, very few fry will survive the attack of these insects in the limited space of the container.

Manuring of Nursery Ponds

The main object of nursing carp fry in specially prepared ponds is to raise a short-term crop of as many healthy fingerlings as possible in the limited area of the pond. As the number of fry that the natural resources of a pond can support until they attain the fingerling size is generally very limited, the pond must be specially prepared to support a much larger number of fry than its size and resources warrant. One important way of doing this is to manure the pond to increase temporarily the supply of food for the fry.

The manures used are either inorganic (chemical fertilizers) or organic. The common inorganic fertilizers are super-phoshpate, ammonium sulphate and sodium nitrate. Very little information regarding the relative utility and dose of these is available in India at present. Inorganic fertilizers tend to increase the production of microscopic floating algae.

In view of the fact that zooplankton are relatively more valuable as a food for early carp fry than phytoplankton, the application of inorganic manure does not seem to be advisable in nursery ponds. The choice, therefore, falls on organic fertilizers.

The commonest and the cheapest available organic manure in the country is cow-dung. Experiments show that cow-dung is one of the best manures for carp nurseries. Heavy manuring with cow-dung in experimental plots or in cement cisterns (dose ranging from 1,836 to 18,357 kg per ha results invariably in the production of a rich crop of Rotifers and Cladocerans and Copepods which appear in thick swarms in the entire column of water within 9 to 12 days. As very few algae are produced at this stage, these is almost a 100 per cent preponderance of animal organisms during a period of about 7 to 10 days, depending on the dose of the manure applied. Appreciable numbers of planktonic algae begin to appear.

Appreciable numbers of planktonic algae bring to appear only when zooplankton swarms decline.

Though microscopic algae are known to be the first link in the food chain in a watery medium, their conspicuous absence in the initial stage of heavy manuring with cow-dung seems to be due to the fact that the animalcules which appear first in the water are capable of directly utilizing in the large quantities of suspended and dissolved organic matter found in the cow-dung.

Heavy manuring with over 3500 kg of cow-dung per ha has not yet been attempted in natural ponds. It is true that an application of 10,000 to 20,000 kg of cow-dung per ha results in a marked increase in plankton populations, but what proportion of it will be zooplankton initially has not yet been definitely ascertained.

In current fish cultural practices the nursery ponds are usually manured after they are cleared of all predatory and other fishes. Cow-dung applied at 1,836 kg per ha produces in about 24 days on an average the maximum quantity of plankton. The dates for manuring the pond and stocking it with fry will depend on the period of maximum production of plankton and the expected dates of availability of the fry. If, for instance, the fry would become available in the second week of July, the ponds should be manured by the 20th of June, but as local conditions vary, no general principles can be laid down as to the interval that should be allowed to lapse between manuring and stocking in successful pond management.

The manure may either be broadcast over the entire bottom if the pond is dry, or over the water surface if it is full, or deposited in small heaps on the sides at selected points. Dumping of manure at one point is inadvisable in view of the fact that the desired heavy crop of plankton in a short period and on a given date is not produced.

The application of raw cow-dung brings about a quicker production of plankton probably because of the fact that the nutrients present therein are more readily available than in cow-dung manure or dry cow-dung.

With the simultaneous addition of green manure (Leaves of plants like *Lantana*) and cow-dung, the plankton produced is proportionately greater, but is very often preceded by an algal bloom in place of the zooplankton swarms obtained with cow dung alone.

Controlling Weeds in Nursery Ponds

In manured nursery pond, weeds particularly the submerged ones, grow very rapidly, utilizing the readily available rich nutrients in the water. The growth is often so quick that a 0.1 acre pond is covered almost completely within a short period of 15 to 20 days. When weeds abound, the plankton in the water is generally very poor. As tender baby fish cannot feed on weeds and as the presence of weeds results in poor production of plankton, it is essential to have them removed from the nursery ponds. As their removal by chemical means adversely affects the plankton, it cannot be resorted to in nursery ponds. The mechanical removal of weeds at regular intervals with human labour or with weed-cutting machines is perhaps the best in shallow nursery ponds.

Marginal weeds offer excellent shelter for predatory insects such as the diving beetle (Cybister). They lay their eggs boring into the stems of those plants. It is, therefore, of importance that marginal weeds and grass, both on the slopes and on the bundhs of ponds, should be trimmed to keep the insects under check.

To sum up, the important operations in preparing nursery ponds are:

(*a*) Removal of marginal, emergent, floating and submerged weeds (macro-vegetation).

(*b*) A regular watch to keep the weeds in check by prompt removal as and when they come up.

(*c*) Eradication of the entire predatory or, otherwise, harmful fish population in the pond by poisoning with Mohan oil cake in suitable doses.

(*d*) Manuring the pond with cow-dung at 1850 kg per ha about 10 days after poisoning and 20-24 days before the expected date for stocking, and

(*e*) Application of mustard oil-soap emulsion, 12 to 24 hours before stocking, to kill predatory aquatic insects, particularly the common back-swimmers (Notonectids) which are the most abundant and, consequently, the most destructive among insect predators.

The Food of Fry

Manuring the pond enhances the production of plankton, the food for the growing baby fish:

The quantity and quality of plankton fluctuate, and a clear idea of these fluctuations in the pond is essential to fix the time for stocking

Figure 4.5: Common Cladocera (Water-Fleas) in Fish Pond–The Choice Food the Tender Carp Fry

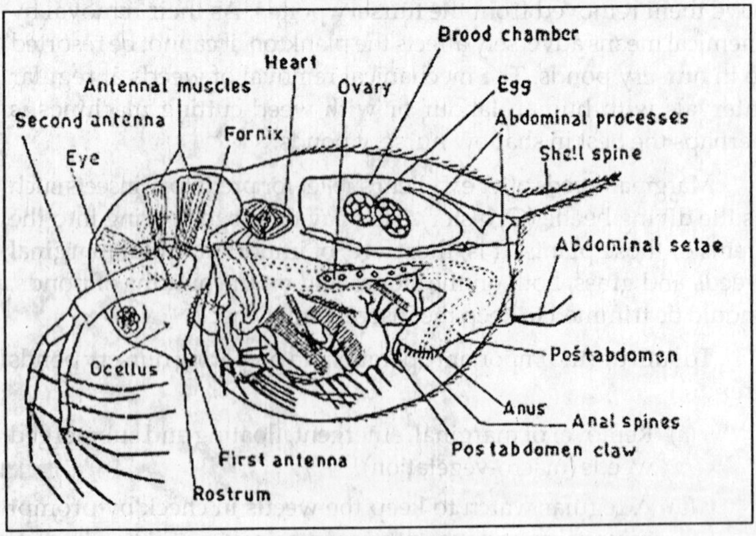

Daphnia similes

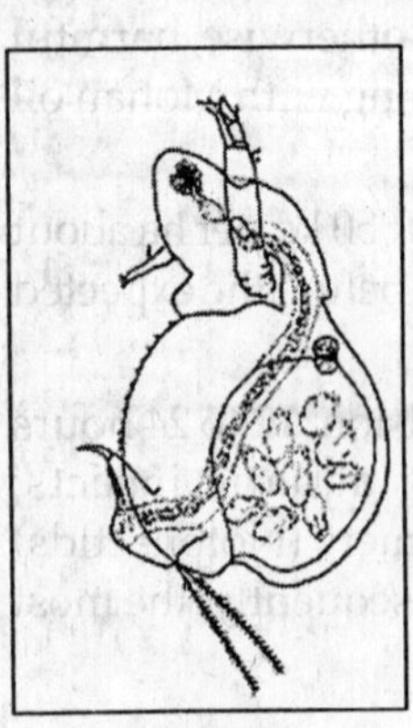

Moina

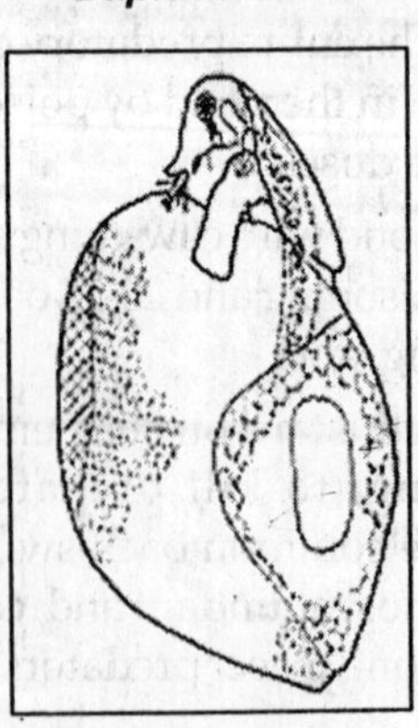

Simocephalus

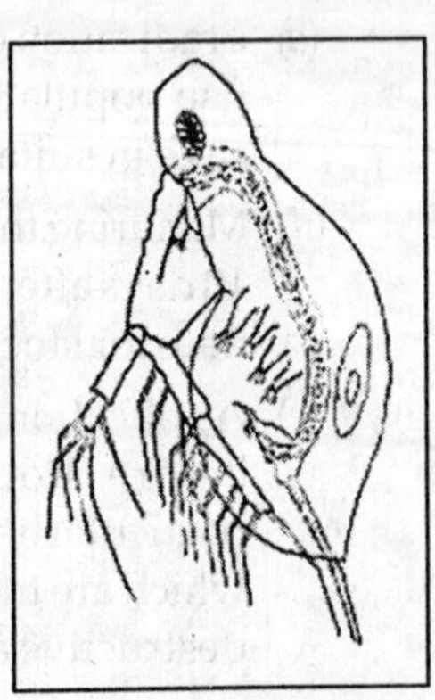

Diaphanosoma

Contd...

Figure 4.5–Contd...

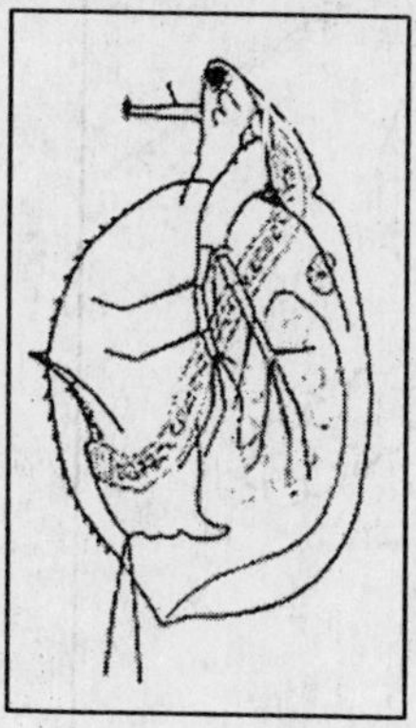

Moinodaphnia

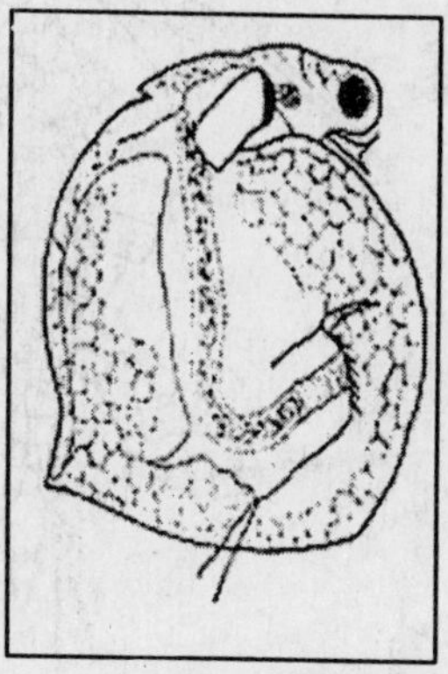

Ceriodaphnia

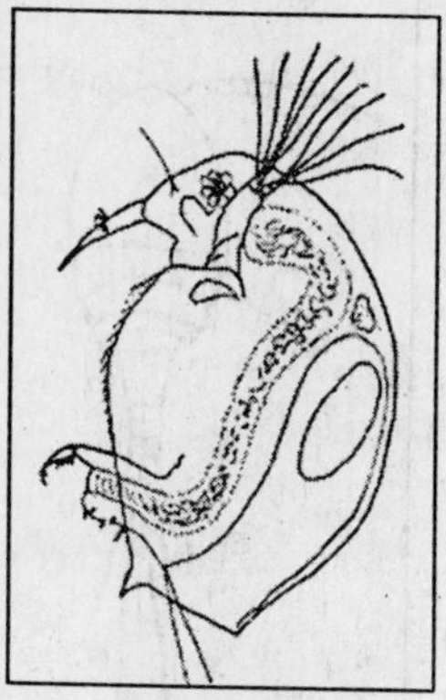

Bosmina

and also the number of fry to be stocked. It is necessary to estimate the plankton quantitatively every week or, preferably, every fourth day after manuring the pond. A small conical net (called plankton-net) made of bolting silk cloth, open at both ends and fixed on to a circular metal frame 22.5 cm to 30.5 cm in diameter and provided with a short handle is used for making collections. A glass tube, 7.5 pr 10 cm long and 2.5 cm in diameter, is tied firmly to the tapering tail-end of the net.

About 50 liters of water taken from 12 different and representative places in the pond area are filtered through the broader mouth-end of the net. Samples of water are collected in a mug of 1 liter capacity by a person standing in the water near the edge of the pond and stretching out his arm towards the centre of the pond.

The surface of the pond is first disturbed slightly to disperse or mix the algal scum, if any, and the mug is held with its mouth downward to about 15.0 cms below the surface of the water, and gradually tilted to collecte water. The plankton-net is held in the left hand with the net portion immersed in the water up to 1.25 cm or 5.0 cm from its mouth, and the water in the mug is gently poured into the net. This process is repeated with water collected from other selected points in the pond to filter 50 liter of water. The net is moved up and down in the water without immersing the mouth so that the water is rapidly filtered and the fish food organisms are washed down into the glass tube tied to it's lower end. The tube is then

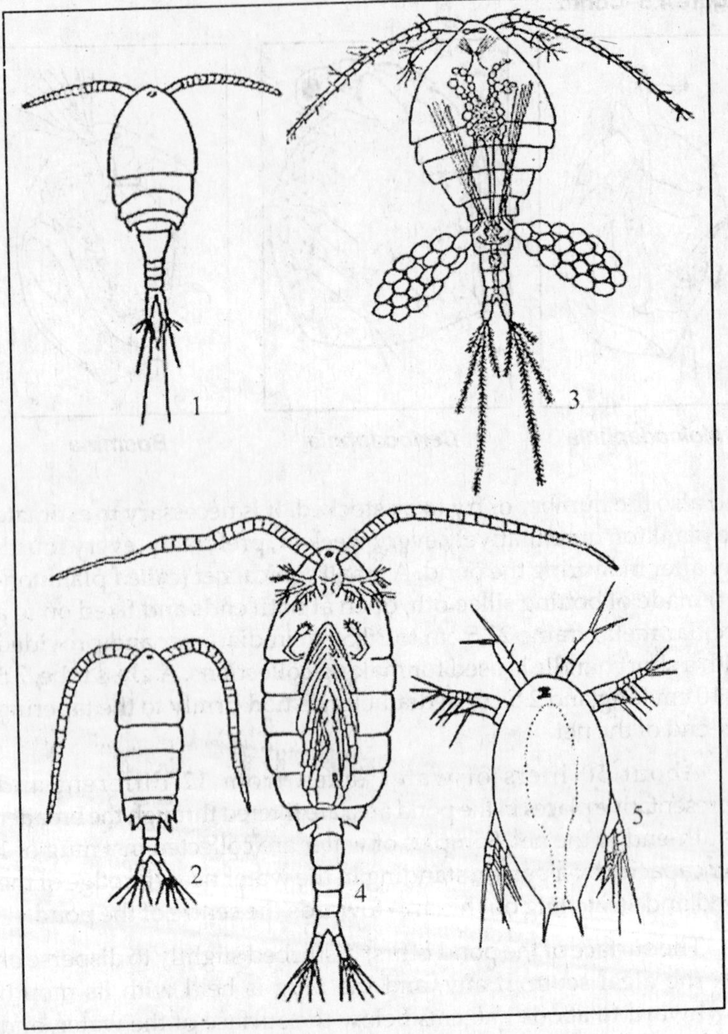

Figure 4.6: Common Copepoda in Fish Pond–The Choice Food the Tender Carp Fry

(1) Cyclopoid copepod; (2) Calanoid copepod; (3) *Cyclops*; (4) *Diaptomus*; (5) *Nauplius* larvae

untied from the net and 15 drops of strong formalin are added to the water in the tube to kill the plankton organisms. Within 15 to 20 minutes of adding the formalin, most organisms settle down at the bottom of the tube, as a sediment.

If this column of plankton sediment is 1/4 to 1/3 cm high from the bottom of the tube, and is found to consist mostly of water-fleas and other animalcules, the water in the pond may be considered to be sufficiently rich in plankton to be stocked at the rate of two to three lakhs of fry per 0.4 ha. The animal or plant nature of the plankton sediment is roughly indicated by a pale brownish or green colour, respectively.

When not required for further study in the laboratory, a rough determination of the quantity and nature of the plankton may be made by the addition of a pinch of powdered common salt which would serve the purpose of quickly sedimenting the plankters equally well as formalin. After noting the approximate height of the sediment in the tube, the collection could be thrown out and the tube used for fresh collections after rinsing.

Nursing Eggs to Hatchings

The hatchery pits with the hapas fixed in position and water running through them should be kept ready for receiving fish eggs for hatching. Whether from bundh-type tanks or from rivers, the eggs collected usually consist of a mixture of several species of major carps, minor carps and minnows, and probably of catfishes as well. With these, there will also be aquatic insects, insect larvae and vegetable debris. The fish eggs, therefore, will have to be sorted out from all these before they are transferred to the hatching pits. On collection, the fish eggs should be examined in shallow flat dishes to carefully separate them out from aquatic insects and the debris.

They should then be transferred to a sieve with mesh 3.0 to 3.5 mm in diameter, kept partly immersed in water and gently moved about as in sieving. Eggs less than 3.5 mm in diameter and which will pass through the sieve may be discarded, as the swollen fertilized eggs of the major carps are larger than these. Those retained on the sieve may now be transferred to the inner cloth tank or hapa made of round-meshed mosquito netting. A number sufficient to cover the bottom in a single layer may be kept in each hapa for hatching.

The eggs thus collected will be in different stages of development, with the embryo not formed in some and fully formed and active in others. They will hatch out within a period varying from 2 to 15 hours after collection. Being slender and small, the tiny

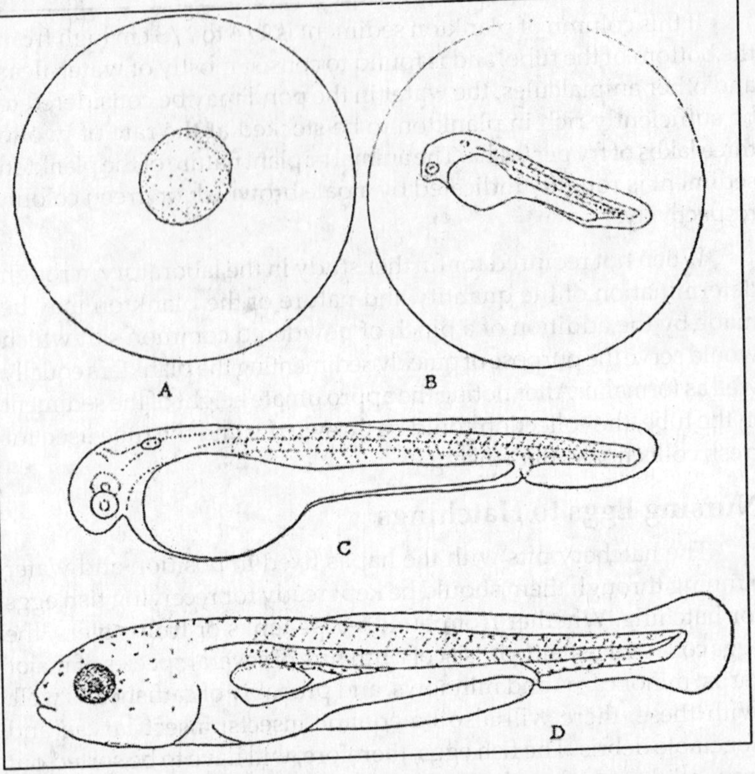

Figure 4.7: Stages in the Development of a Carp

(a) Fertilized egg with the fully swollen egg membrane; (b) fertilized egg with the young fish fairly well developed inside; (c) baby fish just hatched out of the egg; (d) baby fish two days after hatching. All these stages are collected from rivers and bundh-type ponds.

hatchlings pass easily through the meshes of the hapa into the outer cloth tank, while the burst egg membranes are left over in the inner one. About 20 hours later, when all the eggs have hatched out and the hatchlings have passed into the outer tanks or cloth, the inner mosquito-net hapa may be carefully removed along with the egg membranes.

When the hatchlings come out of the eggs with their conspicuous yolk-sac, they are about 4 to 5.5 mm long and subsist on the yolk in the yolk-sac for at least two days; after this they begin to feed on organisms found in the water. As the baby fish in the small hatching pits cannot get much food in the flowing water that passes through, it is essential to transfer them carefully to well-prepared nursery ponds on the third day after the collection of eggs. As the hatchling are kept in hapas in the hatchery pits, they are collected easily in hundies and transported to nursery ponds.

Nursing the Fry (Hatchling and Larvae)

The fry collected from rivers are generally a few hours to two to three days old after hatchling from the egg. Within a few hours to two to three days after collection from the river, they are released into nursery ponds in which their rate of survival is usually very low and fluctuating.

In experiments in laboratory aquaria, over 50 per cent of the tender fry survived when kept in pond water without food for seven days. A much larger percentage (80–96) of the fry survived during the first week when the water in the aquaria was even partially renewed. Daily feeding with freshly collected water-fleas and other animalcules helped to maintain the fry in an excellent condition, contributing to 96 to 100 per cent survival during the first week. In these experiments, predatory organisms and other enemies of carp fry were completely eliminated.

These simple experiments showed that when the enemies of fish fry are totally absent in their habitat, a large scale mortality of fry does not occur during the first few days no matter how starved the fry might be. The experiments also showed that under similar conditions but with a regular supply of artificial or natural fry food, the survival rate is almost cent per cent, except when the fry has suffered injury. If such favourable conditions prevail in the natural nursery pond also, the rate of survival of fry will be high.

The laboratory finding were tested on a miniature field scale, in open cement cisterns, 3.5 m × 1.8 m × 1.0 m in dimensions. Swarms of water-fleas (Cladocerans) and other animalcules (Rotifers) were induced in the water in the cisterns by heavy manuring with cow-dung and the cisterns were kept covered with fine-meshed wire-netting to keep away flying aquatic insects. At the time of stocking with fry, the water in the cisterns was dark, chocolate-coloured and had very low dissolved oxygen (0.8 to 1.4 parts per million only).

The cisterns were fished on the 10th day after stocking. The number of fry stocked ranged from 3,500 to 6,000 per cistern. Mortality was negligible and the survival ranged from 90.7 to 99.3 per cent. These result, besides fully confirming the laboratory findings, showed that

Figure 4.8: Planktonic Algae taken only as Emergency Food by Young Carp Fry

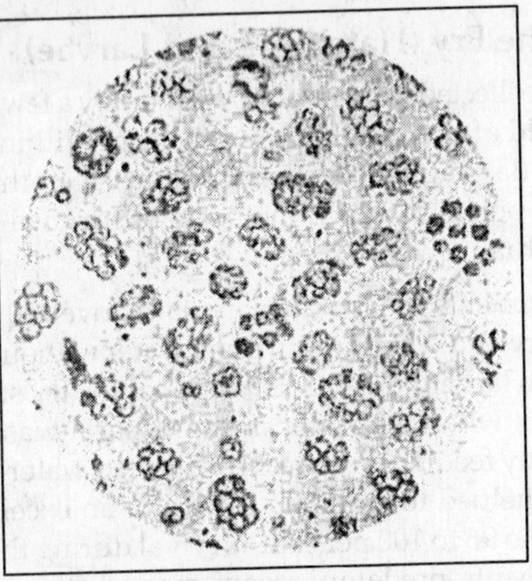

Pandorina

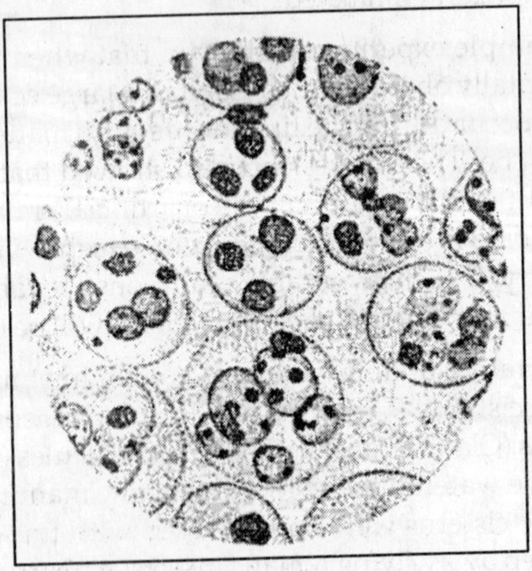

Volvox

Contd...

Figure 4.8–Contd...

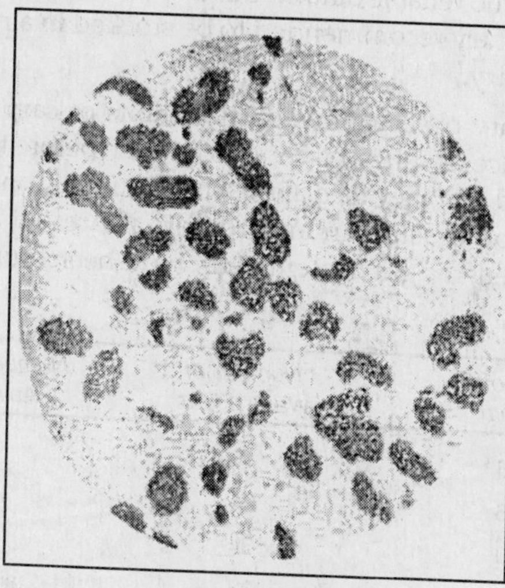

Microcystis

the control of fish fry enemies can successfully be attempted on a larger scale than in the laboratory. It also showed that will the provision of ample food, the tender fry would tolerate even dangerous living conditions, particularly a low dissolved oxygen content.

Further experiments showed that the maximum growth and maximum survival of fry are obtained when they are fed on zooplankton consisting mainly of water-fleas. The growth and survival of fry were considerably low when they were given only algae food.

Experiments have also revealed that very young carp fry, with their relatively short, straight intestine, characteristic of carnivorous animals in general, are incapable of digesting atleast some of the planktonic algae either because the algae are resistant to the digestive secretions if any, in the gut, or because these items are not retained in the gut sufficiently long to be digested. On the other hand, the water-fleas and animalcules consumed are digested very quickly, judging from the remains voided through the vent within the hours of feeding on them.

The number of fry to be stocked in a pond depends both on the feeding capacity of fry and the quantity of plankton available in it.

As at present, no reliable data on this important aspect are available, the number of fry recommended to be stocked in a pond would be largely arbitrary.

Laboratory observations on the habits of carp fry show that they commence feeding at a very early, even before the yolk is fully absorbed, and that they feed almost continuously on their favourite food, plankton crustacean and rotifers if present in abundance. Under full feed, the gut content of fry have been found to consist of the following.

Length of Fry in mm	Number of Water Fleas	Number of Wheel Animalcules
6.8	10	1
7.6	21	1
8.0	34	2
9.8	23	14+2 algal filaments

A single carp fry 6.5 to 7.2 mm long consumes within an hour 3 to 34 water-fleas, the actual number depending on the size of the fry and the stage of absorption of its yolk. Fry of the same size take in about 130 to 150 water-fleas within period will depend on their availability in the environment and also on the number of fry feeding there.

When the density of zooplankton under field conditions was taken as the criterion for stocking nursery ponds with carp fry, the rate of survival of fry was seen to be very satisfactory in several ponds but poor in others. Ponds with satisfactory rates of survival of fry contained abundant zooplankton at the time of stocking, and very little sign of planktonic algae bloom during the entire period of rearing. Ponds having algal blooms at the time of stocking of during the period of rearing give only medium rates of survival of the fry stocked.

As no measures for the control of predatory insects are taken in the ponds, the high rates of survival (even up to 90 per cent) obtained in some of them can be said to be due alone to be deliberate choice at the time for stocking. When the insect population is poor, larger number of fry survive, and conversely, when this population is heavy, the majority of the fry do not have a chance to survive even if the food supply is plentiful in the water. Ponds in which the insect population is controlled by suitable treatment given decidedly better yields of fish than ponds not so treated.

Supplementary Feeding

It is difficult to maintain a continued high level of production of zooplankton for more than a few days in ponds. Even if the ponds are rich in zooplankton at the time of stocking, they suffer depletion within two to three days of introducing the fry. It is therefore, difficult to maintain the same density of zooplankton with a heavy population of fry in them. In certain ponds, however, the reproductive capacity of plankton at the time of stocking is so great that even after a large number have been consumed by the fry, the volume of plankton remains undiminished in the first few days. If the insect are controlled in such ponds, the rate of survival of fry can be maintained at the high level.

When the plankton density in the pond goes down rapidly, there is great competition for food amongst the fry which have fewer chances of survival and growth, if the food resources are not maintained at a high level. Besides, there are at present certain practical difficulties in maintaining a high level of zooplankton incidence in ponds. Manuring alone may not increase the production of zooplankton, as algae blooms often intervene, affecting their growth. Similarly, the duration in which manuring takes effect with reference to production of plankton can also never be forecast. These uncertainties, together with the fact the supply of fry is restricted to a few place and to a very short period, not exceeding a couple of weeks in the year, make it difficult to have satisfactory fish cultivation.

Laboratory experiments have shown that the tender carp fry feed avidly as soon as they are able to do so, on artificial foods such as rice bran and groundnut oilcake or a mixture of the two.

The experiments show that carp fry grow decidedly better when equal quantities of zooplankton and artificial food, rather than an excess of artificial food alone, are given and that they consume more of the zooplankton than of the supplementary diet in a given time. As such, an excess of the zooplankton over artificial food helps their growth better than an excess of artificial food over zooplankton. The fact that for the first two days the fry make hardly any use of the supplementary food will have some bearing on their condition and capacity to withstand enemies in the water.

It is, therefore, clear that for the successful rearing of fry, the nursery pond should be rich in zooplankton at the time of stocking, and that from the second of third day after stocking, the natural zooplankton food should be supplemented by the supplementary food such as rice bran and oilseed cake.

Under field conditions, the fry have to be fed once or twice a day with rice bran, and/or oilcakes (groundnut, coconut and mustard). These are more easily assimilated by the delicate fry if they are in a finely powdered state, when they will also remain suspended in the column of water for a much longer period. The rice bran should be passed through a fine-meshed sieve to remove the husk. Similarly, the powdered oilcakes should also be sieved to ensure uniformity in the size of particles. In a series of experiments with coconut, mustard and groundnut oilcakes and rice bran, which gave very satisfactory results, the daily dose of artificial feed given ranged as follows :

- ☆ *First five days after stocking*: Daily, twice the weight of fry at the time of stocking
- ☆ *6th to the 10th day after stocking*: Daily, thrice the weight of fry at the time of stocking
- ☆ *11th to the 15th day after stocking*: Daily, four times the weight of fry at the time of stocking.

With mustard oilcake, however, the dose should be lower. Start with the weight of the cake almost equal to the weight of the fry and double it in two stages. In actual practice, it is difficult to weigh of a sufficient number of fry to calculate the exact weigh of food to be given to them. Normally, a mixed collection of a lakh carp fry (consisting of sizes from 6 to 6.5 mm in length and each weighing, or an average, 0.0014 gm) which will be contained in a cup will weigh about 140 gms. From this, the quantity of food required for any given number of fry can be easily calculated. The following proportions of artificial food in terms of the volume of fry would be helpful.

Volume of Fry (in 'bati')	Approximate Volume (in 'batis') of Artificial Food			
	Rice Bran	Groundnut Oilcake	Coconut Oilcake	Mustard Oilcake
1	2.0	1.25	2.0	1.5

Very little experimental work has been done on supplementary feeding of carp fry in India. With regular daily feeding with selected items of supplementary food at doses varying from one to four time the initial weight of the fry, the average production of early fingerlings for the first 15 days of rearing has been as follows :

Supplementary Food Provided	*Number of Ponds Used for Rearing*	*Percentage of Survival Range*	*Average*	*Average Production Number per Hectare*
Rice bran	4	47.0 to 99.0	77.2	3,85,930
Groundnut oilcake	5	13.2 to 71.8	36.6	1,93,070
Mustard oilcake	4	4.5 to 51.0	28.2	1,41,020
Coconut oilcake	2	47.8 to 61.0	54.4	2,72,120

The highest rate of survival of fry was in the pond to which rice bran was added daily at twice the initial weight of the fry. After 15 days of rearing, the pond was thoroughly fished and the early fingerlings removed. The pond was re-stocked with another lot of early fry and fed daily as before. During the second 15 days of rearing also the rate of fry survival was extremely satisfactory, though the pond was not manured afresh after the first fry crop was removed.

Within a period of five weeks, a total of 94,000 advanced fry and early fingerlings was raised in two crops in a 0.04 ha pond. With the application of the well-tried methods of nursery preparation, stocking and feeding the fry, such a high rate of production of advanced fry and early fingerlings high rate of production of advanced fry and early fingerlings should make fish culture a really lucrative undertaking. In the present instance, the yield of fry 1.9 cm in length in both the crops should fetch in all Rs. 400 at a nominal rate of rupee five per 1000 fry, making allowance for a mortality of as many as 14,000 (roughly 15 per cent) during collection and transport.

Nursing Advance Fry

The preparation of rearing ponds is no less essential, though less elaborate, than nursery ponds. Here, the essential steps are elimination of all fish from the pond and manuring it prior to stocking with fingerlings.

The feeding habits of the fingerlings are different from those of the young fry, though Catla in the fingerling stage also takes in large quantities of water-fleas. In weedy ponds, the growth of Catla is very slow. Although Rohu and Mrigal feed largely on decaying weeds and plankton, their growth is usually unsatisfactory in pond overgrown with weeds. The overgrowth of weeds should, therefore, be controlled in rearing ponds. A sparse distribution of submerged weeds-altogether about 1/8th the area of the pond-without appreciably impoverishing the water, may be ample for a healthy growth of Rohu and Mrigal. However, no reliable data are available at present on the extent to which weeds are used as food by fingerlings of major carps.

The mortality rate of fingerlings is relatively much lower than that in the fry stage. Fingerlings transported over long distances are liable to be weakened for want of food or are liable to be injured, and suffer heavy mortality in ponds. But those exceeding 2.5 cm in length and in a healthy condition suffer less mortality. In seven months of stocking healthy fingerlings, the mortality has been found to vary from zero to five per cent in Catla, and 5 to 10 per cent in Rohu and Mrigal. From 6,000 to 8,000 advanced fry or early fingerlings (1.9 cm to 3.8 cm) may be maintained in a healthy condition in a 0.04 ha pond for a period of about two months if regularly fed with groundnut oilcake. At the end of this period they will have grown to a length of 5 to 6 cm. But the poor growth attained in this period despite artificial feeding would indicate that the stocking rate is too high for the area of the pond. It is, therefore, necessary to thin out the stock if, on periodic examination, the growth registered is too little for the area of the pond.

One of the most important aspects of successful rearing of fingerlings and adult fish is the correct stocking of the ponds.

Nursery Ponds

As the nursery ponds are heavily stocked and the food requirements of the growing fry increase tremendously, the stock has to be transferred to other ponds. Advanced fry a early fingerlings are generally released is rearing ponds in which they are kept till they attain the advanced fingerling or yearling size. To secure optimum growth and production in carps in such rearing ponds, the fry or fingerlings should be sorted out according to their size and stocked in known combination.

Stocking Fry and Fingerlings

How many fish are to be stocked in a pond is an off-repeated question. For want of reliable data, no satisfactory answer can be given for it at present. As the growth of the different species of fishes, like their food and feeding habits, varies, the rate of production of fish in a given pond will depend on the species of fish selected and the proportions in which they are introduced so as to make the fullest use of all the food available in the water.

As single species farming helps in the utilization of only a few of the several food resources of the pond, a combination of two or more species which can utilize all the available food in the pond should be chosen as such a mixed farming is an economic proposition. It is important, therefore, that the species chosen should be mutually compatible, that is to say they should neither compete for food nor prey upon one another.

The proportions of these compatible species should be based upon the estimated stock of the various types of available and renewable food resources in the pond. As the stock of available food varies from place to place depending on soil conditions, rainfall, temperature, depth of water, vegetation, etc., in the pond, the proportions will naturally vary for different ponds. As a correct stocking based on these several factors will alone ensure a satisfactory based on these several factors will alone ensure a satisfactory and economical rate of growth, care must be exercised in selecting the species and calculating their number and proportions for each individual pond.

As has already been stated, nursery ponds (ponds for stocking tiny fry) and rearing ponds (ponds for rearing fingerlings) have to be carefully prepared before they are stocked. A similar but less elaborate treatment is necessary for stocking pond with yearlings and adults as well. Eradication of predators, manuring and artificial feeding will ensure high returns in stocking tanks.

As the species are indistinguishable in the nursery stage, the tender fry of several species are all stocked together. While there are no hard and fast rules regarding the number of fry to be stocked, the number should depend upon the available nutrients in the pond or its food-producing capacity. It has been suggested that small nursery units be stocked with tiny fry at the maximum rate of over 2,22,390 per hectare.

In the nursery ponds, the highest rate of stocking in general rearing experiments with artificial feeding was 12,35,500 fry per ha of water surface. This resulted, in some cases, in over 90 per cent survival in the course of the first 15 days. A much heavier rate of stocking was possible in cement cisterns (3.5 m × 1.8 m × 0.75 m) with about 5,600 fry of which over 5,400 survived within the first 10 days when the stock was finned out. There was no artificial feeding. The rate of stocking works out to approximately 84 lakh per ha of water surface. It should not be very difficult to increase this number with artificial feeding. From theoretical considerations, such a heavy stocking in natural nursery ponds should be a distinct possibility.

Repeated laboratory experiments have shown that regular feeding with zooplankton produced excellent results in growth. Thus, a 100 tiny fry stocked in three liters of water in a glass jar were successfully reared to over 2.5 cm size within 15 days.

The food requirement of the fry increase with their growth. As the nursery pond cannot cope up with such heavy demands for food, the grown up fry should be transferred to other ponds which are free from all other fish and are manured heavily. In these rearing ponds also, selective stocking of fry will not be practicable, as individuals 1.9 cm to 2.5 cm long are usually indistinguishable from one another the repeated handling in sorting at the stage results in fatal injury to a large number of them.

It has been stated that a pond 15.25 m × 15.25 m × 3.0 m will support 5,080 cm long fry for four months. Such deep ponds are not required for rearing advanced fry. Taking the area alone, the rate of stocking should be about 86,485 per ha, although a higher rate of stocking (about 1,48,260 per ha) has also been suggested.

In experimental rearing ponds 1.2 to 1.5 m deep, 1.9 to 2.5 cm long fry can be stocked at the rate of about 24,710 per ha, when a majority of them will survive and grow to about 7.5 to 10.0 cm in length in the course of about three months. No supplementary feeding need be resorted to in this case. If the pond is more heavily stocked, the percentage of survival generally goes down. With daily artificial feeding with finely powdered groundnut oil-cake, the stocking rate may be raised to about 1,97,680 per ha, with the survival rate exceeding 90 per cent during a period of about six weeks in which they attain an average length of about 5.0 cm. For further satisfactory rearing in the same pond, the stock has to be thinned out to about half.

The necessity for selective stocking even at the early fry stage is indicated by the fact that in heavily stocked ponds Catla does not grow satisfactorily in spite of artificial feeding. Though it registers the fastest growth among the major carps, if its population in a given pond happens to exceed its stockable capacity, its growth is very much retarded, particularly since it makes the least use of the supplementary food provided, with its upturned mouths adapted to plankton feeding. Laboratory experiments have confirmed that Catla takes a very limited quantity of supplementary food in the form of oilcakes.

Only advanced fingerlings are released in stocking ponds. The number and kind of such fingerlings should be decided on the basis of the size of the pond, its food resources and the species of fish available for stocking. The stocking ponds, like the nurseries and rearing ponds, should also be cleared of all predatory and undesirable fishes. Old tanks with excess of bottom silt should be dewatered and partially desilted and all overgrowth of macro-vegatation also removed. As in the case of the nursery of rearing ponds, the stocking pond should be manured, but in smaller doses and at monthly intervals.

Compatible Species

Before stocking, the food resources in a pond should be assessed. Plankton, macro-vegetation, weed-dwelling fauna, bottom fauna and organic detritus are the main natural food resources in the pond. The relative approximate abundance of these may be judged by random collections from known areas and estimating their volume or weight. This information, together with that available on the food and feeding habits of the different species of fishes will enable a judicious selection of the right species for the pond.

Catla feed largely on zooplankton even after the fingerling stage. The species of the genera *Labeo* and *Cirrhina* are bottom or column feeders taking in large quantities of bottom mud and debris besides rotting aquatic plants.

A satisfactory combination of fishes for a pond should include a plankton-feeding species (column-surface feeder), a detritus of bottom fauna feeder (column-bottom feeder) and a macro-vegetation feeder. As already stated, the proportions of three species will depend on the relative abundance of the different types of food present. If algal blooms are the dominants as in most religious institutional tanks. If zooplankton is rich, Catla will find the conditions good. When there is luxuriant vegetation, Gourami should be introduced. Rohu and Mrigal grow well in ponds with abundance of decaying vegetable matter, including decaying phytoplankton.

Mixed farming of carps is practiced in different countries, particularly in China, where a great variety of stocking has been evolved for ponds in the different regions of the country. In India also, the major carps Catla, Rohu and Mrigal, often with a varying number of Kalbasu and Bata, are stocking together in ponds in West Bengal, Bihar, Orissa, Punjab and Assam.

No fixed proportions for the different species have, however, been evolved, though a combination of 30 per cent Catla, 30 per cent Rohu and 40 per cent Mrigal is reported to be in vogue in certain Bengal fish farms. In view of the fact that most West Bengal fish farmers do not sort the fry of fingerlings into species, the above proportions are not adhered to in actual practice. Reliable data on the proportions of species and the density of stocking for Indian waters are wanting.

Introducing Predators

Predators (murrels, etc) and non-predatory species (carps, etc.) do not go together in the same pond. Catla, Rohu and Mrigal grown in stocking ponds do not seem to feed on aquatic insects and control them. Some of the smaller predatory fishes like *Anabas testudineus, Channa punctatus, Heteropnuestes fossilis, Clarias batrachus* and *Glossogobius giuris,* which are predominantly insectivorous in habit, do not cause appreciable harm to other fishes. Besides, because of their relatively small size, they are generally incapable of doing any harm to grown up fingerlings (12.5 to 15.0 cm) of major carps.

It has been stated that some fishes need exercise for their healthy growth which is provided by their being chased by predators deliberately introduced into ponds. Whether such exercise is really necessary and whether or not the introduced predators take a toll of the carps cannot be definitely said. However, if a small predator which does not harm the carps is introduced, it may increase the total yield of fish to some extent, feeding on the insect population in the pond. Carp fingerlings below a particular length (say, 12.5 cm) should not then be stocked. The predator should also be stocked preferably at 7.5 cm to 10.0 cm fingerlings size.

The high market value of *A. testudineus* as a food fish, its small size and its pronounced insectivorous tendencies make this fish an obvious choice for introduction into carp ponds. *H. fossilis* and *C. batrachus* may also be included in this category, though the latter grows to a fairly large size. *Channa punctatus* and *G. giuris* have relatively larger mouth and should, therefore, be utilized with care.

Even a full-grown medium-sized predator like *Channa striatus* would be incapable of causing appreciable harm to a stock of carps of minimum length of 30 cm.

Ponds Management and Harvesting

After the pond is stocked, the condition of the fish should be periodically given a check up to ensure that they are thriving well.

If weeds overgrow, the excess of growth should be periodically removed. One or two small patches of submerged or floating weeds occupying about one-tenth of the water area, however, may be maintained in the pond without harm. Weeds in moderate numbers not only serve as anchorage for snails and insects, but also as a source of natural decaying vegetable debris. Floating vegetation like duckweeds should never be allowed to cover up the entire water surface.

Unless the ponds are shallow (depth of water below 1 m), shade is not essential. Too much shade is not desirable, and when trees are present on embankments, care should be taken to periodically remove the leaves falling into the water.

Netting the pond once a fortnight, or at least once a month, with a drag-net will provide sufficient run and exercise for the fish besides enabling us to examine them. If they have an emaciated or weak appearance, part of the stock should be removed to another pond. Unwanted fish minnows should be removed at every netting.

During the summer months, with the level of water falling steadily in the pond, the fish are prone to infection by parasites. Periodic netting will enable timely detection of such sites. Periodic netting will enable timely detection of such infection, if present and the taking of suitable remedial measures.

If pond have inlets or outlets or both connecting them with other ponds or water sources, these should be fully secured by wooden or wire-net doors or screen so that the fish stock from the pond do not escape and no unwanted fish get into the pond. Live fishes like murrels and the climbing perch from neighbouring waters cannot, however, be prevented from crawling over moist land into the pond. By cutting the outer side of the embankment vertically to a height of 45 cm at the top of by fixing a fine wire net fencing about 30 cm high on the top of the embankment, or by constructing a cement parapet 45 to 60 cm high all round, unwanted fishes and fish enemies like snakes and tortoises, can be prevented from gaining entry into the pond. Fish-eating birds like cormorants, kingfishers and gulls should either be scared away or their number reduced by periodical shooting.

As the growth of fish varies, individuals of the same size and age belonging to the same species show different rates of growth, though all of them grow in the same pond under the same environmental conditions. If there is very considerable difference in the size of individuals of the same species in a pond, it shows that the smaller fish are not growing properly. When the ponds are periodically netted, the fish should be sorted out according to size and the bigger ones removed to another pond, thus giving a chance to the smaller ones to grow quicker.

Under normal conditions, Indian major carps have their maximum growth in length in the first year of their life, but the maximum growth in weight is probably in the second or third year. When such growth in weight has been attained, the fish should be removed, enabling the smaller fish to grow rapidly. This is a fundamental principle in fish farming where the harvesting of the fully ripe fish crop, as in the case of agricultural crops, should never be postponed.

When seasonal waters reach their lowest level in summer, the fish found in them should be transferred to deeper perennial ponds, if available, for further growth and fattening. Otherwise, the yearlings may be marketed. When the ponds under stocked, the young fish have a chance to grow rapidly and attain a marketable size even by the end of the first year.

Ponds should not be netted during the breeding season or soon after fishes have bred in them. This will result in injury to and mortality of their progeny, apart from scaring way the breeders from the breeding grounds to which they may not migrate again to breed.

The best season for harvesting a pond is summer when netting is easy because of the low level of water, but the date and time of netting should be so arranged as to coincide with the demand in the market. Netting is best carried out early in the morning before the water gets heated up, so that the residuary fishes in the pond are not adversely affected. The use of a large-meshed net (2.5 cm) helps catch only the larger fish, leaving the smaller ones uninjured in the process of netting and less exposed to the chances of catching parasitic or other infection.

Pond Breeding of Carps

As far as is known at present, in confined waters a Indian major carps breed only in the bundh-type tanks or ponds of the Midnapore and Bankura district of West Bengal. Breeders are generally present in wet bundhs. In dry bundhs, they are introduced from ponds when rain water accumulates in them in the rainy season. In most bundhs, the carps spawn in the shallow areas adjoining the outlet. When the water level in the bundh suddenly rises as a result of the inrush of water from the catchment area flooding the spawning grounds, the breeders play about and breed in the shallows. The water from the bundhs may be overflowing through the outlet and the time of spawning. However, in certain years, the fish have been observed to breed even when the usually spawning ground was not at all flooded and no water was flowing out of the bundh through the outlet. The factors which induce the fish to spawn are not yet clearly known.

As in the existing practice of carp-breeding in bundhs no precaution of eliminating unwanted fishes including the predatory species is taken, the eggs or fry obtained from the bundhs (particularly the wet ones) are no less mixed in quality than those obtained from the rivers.

The eggs collected from bundhs are hatched in adjoining hapa-pits and sold on the spot within two to three days of hatching. The owner generally does not keep the fry in the hapas for a longer period than is essential to avoid heavy mortality, but this compels him to sell his stock of fry at very low rates. The loss incurred thus can be minimized or eliminated in two ways :

1. By keeping heavily manured pits ready at hand for maintaining a regular supply of zooplankton with which the fry could be fed from the second day after hatching. A 2 m × 1m × 1m manure pit with at least 0.75 m of water in it, manured with 18 to 22.5 kg of cow-dung would be enough to produce, in about a week's time, all the zooplankton necessary to fed about 5,000 fry for 8 to 10 days.

   An adequate supply of zooplankton is judged by the brown colour of the water. The pit should be manured a week before the expected date of spawning of fish. The number of such pits could be increased if required.

   The fry should be fed from the second day after hatching to prevent mortality.

2. By maintaining, as detailed already, previously prepared miniature nursery ponds into which the hatchlings could be reared and kept ready for sale as per demand.

In the breeding of carps in bundhs, it is necessary to estimate the exact number of breeder of both sexes of different species present in them and maintain a stock of only known proportions of selected species. Improved methods of egg and seed collection are also likely to contribute to increased fish seed sources for us.

The common carp (*Cyprinus carpio*) breeds in ordinary ponds with plenty of submerged vegetation to which the eggs water attached.

Live Fishes

The live fishes, unlike the carps, are capable of directly breathing atmospheric air, and have very often to come up to the surface of water to do so. Being thus practically independent of the oxygen fluctuations in the water, the live fishes are capable of tolerating even foul waters in which ordinarily no other fish will thrive. The common live fishes of cultural value in the country are the following.

- ☆ Gourami, *Ospronemus gorami.*
- ☆ The Large Murrel, *Channa maurlius*, found throughout India.
- ☆ The Common Murrel, *Channa striatus*, found throughout India.

- ☆ The Koravai Murrel, *Channa punctatus*, found throughout India.
- ☆ The murrel, *Channa stewarti*, found in northern Bengal and Assam.
- ☆ The Climbing Perch, *Anabas testudineus*, found throughout India.
- ☆ The Singhi, *Heteropnuestes fossilis*, found throughout India.
- ☆ The Magur, *Clarias batrachus*, found throughout India.

All these fishes, excepting *Gourami*, are predators breeding in ponds. Under natural conditions, some of them breed in rivers also, but ponds, swamps and lakes constitute the main seed sources.

As *Gourami* feeds predominantly on macro-vegetation, it should be stocked in weedy waters. Being a perch with strong spines on the dorsal and ventral aspects of the body, advanced fingerlings, 15.24 cms and over in length, are capable of holding their own against medium-sized predators and can, therefore, be stocked even in ponds from which complete eradication of predators is not possible. As it does not directly compete for food with the major carps, *Gourami* could also from a supplementary stock in carp ponds. Being a nest-builder, breeding facilities in the form of submerged and emergent vegetation should be provided for their satisfactory propagation. In ponds devoid of weeds, objects such as dried grass and coconut palm leaf fibers should be introduced to enable the fish to build its nest

Gourami guards the nest till the eggs hatch out. Once the young ones come out of the nest, the parents do not attempt to protect the brood. At this stage, the young ones, which are still more or less helpless, often fall easy victims to the numerous enemies in the water which include predatory insects, minnows, gobys and all other fish in the pond including their parents. The survival rate of young ones is, therefore, limited. To improve the survival rate, one of the two methods described below can be adopted.

1. Special breeding ponds, small and shallow with a constant level of water and an abundance of submerged vegetation should be selected and all other fish removed. After heavy manuring with cow-dung, the ponds may be stocked with a known number of breeders. The males and

females are distinguished by their external features; the male with brighter colour, a smooth knob-like prominence at the nape and rather thick lips, and the female without them are with rather thin lips. When these fish breed, the young ones will be comparatively safe from enemies but the large number of insects and insect larvae in the water will still take a toll of them. Besides, the young fish generally require water-fleas and animalcules (zooplankton) as their main food which can be produced only in very limited number in the pond, as manuring has little effect on their production because of the abundance of weeds.

2. In view of these limitations, an alternative method is to have a few cement cisterns about 4m × 2m × 1m. The breeders are introduced into a weedy pond where careful watch is kept over their nest-building activity. When the nests are completed, the cisterns should be filled with water to a depth of 0.75 m and immediately manured by adding about 27 to 32 kg of cow-dung in each. The cisterns should be kept covered by fine-meshed wire-netting on a suitable frame so that the flying aquatic insects will, to a large extent, be prevented from entering the cistern.

Within 10 to 12 days of manuring, millions of water-fleas and other animalcules are produced in the water, by which time the eggs will hatch out in the nest. Within the next two to three days, *i.e.*, before the young ones go out of the nest, they should be carefully scooped out and transferred to the cement cisterns. With an abundance of the desired food, they grow quickly, the mortality is very low and the fry attain the early fingerling length of about 2.5 cm within a month. The hatchlings from two or three nests may be kept in a cistern. When they grow to about 2.5 cm in length they should be transferred to small natural nursery ponds, also manured heavily beforehand with cow-dung or stable refuse. After rearing them to 5.0 to 7.5 cm, they may be transferred to the weedy waters for growing.

Fish Enemies

The enemies of fish in ponds belong to diverse groups of the animal kingdom as follows:

1. The common aquatic plant or bladderwort (*Utricularia* spp.) which traps and kills small fish fry.

2. A variety of aquatic insects, including the more common water-boatman, water-scorpion, back-swimmer, diving beetle, giant water-bug and dargon-fly and their nymphs, which attack and eat fish fry and even early fingerlings.
3. Minor carps, carp minnows and other miscellaneous small fish found in ponds, feeding on the tender carp fry.
4. Predatory fishes, destructive to carps at all stages.
5. Frogs and tadpoles. The latter though known to be herbivorous, have been found to have eaten carp fry.
6. Water-snakes and tortoises, harmful mainly to fingerlings.
7. Cormorants, gulls, king-fishers, kites, crows, herons, storks, duck, etc. Fish-eating birds destroy large numbers of even one-year old fishes.
8. The common otter is destructive to large fish.

Parasitization of freshwater fishes and frequent mortality among them prevail particularly during the summer months when the water level is low and stagnating. There are several kinds of fish parasites which affect fishes of all kinds resulting in their death. The fry and fingerlings of carp which suffer from frequent handling during collection, conditioning and transport are more liable to be infected than adult fish.

Fin rot is a common disease in fishes caused by bacteria and the water-mould *Saprolegnia*. Infection by bacteria can be detected by the whitish opaqueness of the infected part while fungal infection gives the infected part of white fluffy appearance. Roughly handled and bruised fish are more prone to infection by them. The fins are usually affected, their margins being gradually eaten off, hamphering normal movements. Acute cases in which the entire tail fin is eaten off, spells danger to the hind region of the body which swells, resulting in the death of the fish.

The mould called *Branchiomyces* causes gill rot by obstructing the veins in the gill filaments. If the infection is limited to a few specimens, their treatment for 5 to 10 minutes in a bath of three to five per cent salt solution will stop the infection. If the infection due to bacteria, a similar bath in a weak potassium permanganate solution (five parts in a million parts of water) gives good results.

Injured young fry of carps kept in hapas or pits catch the bacterial infection quickly. Preventive measures such as careful handling during collection, reducing the number of fry in a hapa and keeping them in gently flowing water till they are released in ponds, go a long way in keeping down infection, but treatment of fry with a solution of potassium permanganate or common salt arrests its spread.

If there is a large scale infection in ponds the fishes should be transferred to clear ponds, and the water in the infected ponds drained completely and the pond bottom treated with lime.

As protozoan parasites (*Myxosporidians*), which often form yellowish or pale cysts on the body or on the gills, are not easily killed by saline or permanganate baths, the infected fishes are best removed and destroyed.

The almost invisible worm *Gyrodactylus* is a common parasite or carps in ponds, particularly Rohu (*Labeo rohita*). By attaching itself with its long hooks to the gill, the skin and other parts of the body, it causes irritation to the fish. A five-minute bath in a weak solution of acetic acid (one part of acid in 500 to 1,000 parts of water) followed by another in a three per cent salt solution is, however, sufficient to control it.

Carp lice (*Argulus*) and the allied *Lernaea* and *Ergasilus* belonging to the Crustacea are an important group of fish parasites. *Argulus* attaches itself to the body of the fish by means of suckers and hooks, but it can also swim freely in water. Larger-scale infection of *Argulus* in ponds and rivers is not uncommon, with some acute cases of infection in ponds known in which over 2,000 specimens have been found on a single Rohu. The infection results in the slow sapping of vitality of the fish to the point of death.

A few wooden or bamboo poles fixed in different place in the pond are said to help the infected fish to get rid of the parasites by rubbing their body against them. In case of a large-scale infection, application of a weak alcoholic solution of 'Gammexane' (1½ to two parts of 'Gammexane' for a million parts of water) to the pond has checked the mortality of infected fishes.

Lernaea is a minute rod-like of filament-like external parasite sticking out from the angle of the shoulder-fin and body on the tail or other fins, from the eye socket, from the corner of the mouth and

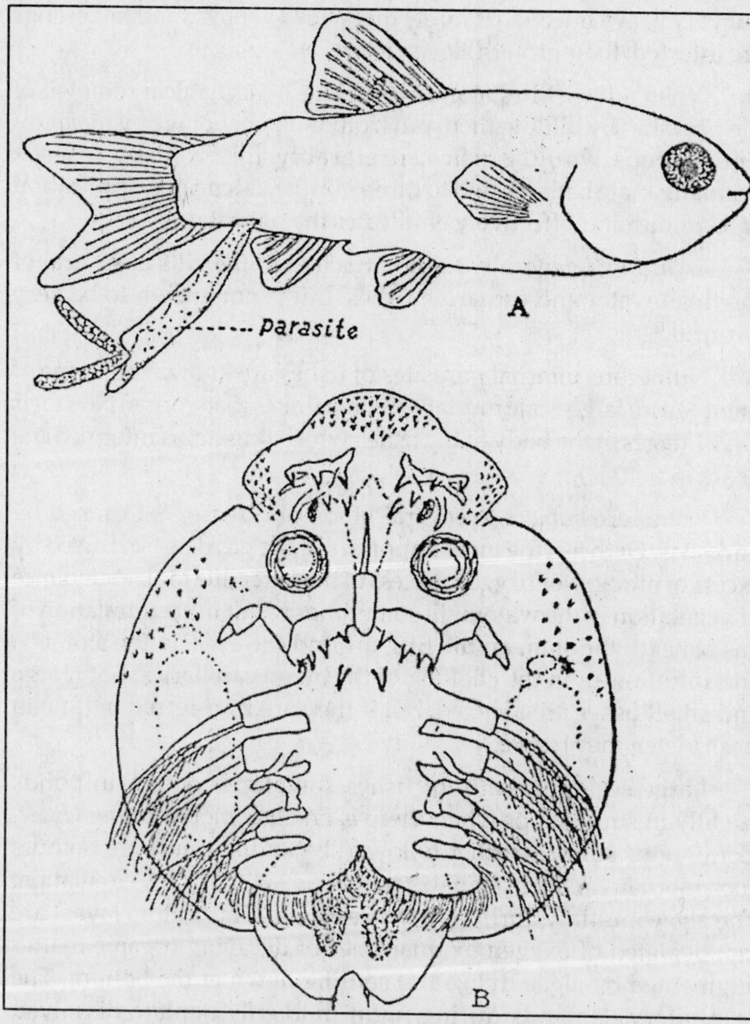

Figure 4.9: Common Fish Parasites

(A) A 14 mm. long fry of the minor Barbel Barbus stigma, infected by a parasite (*Lernaea*), seen sticking out from the distal portion of the body which is consequently swollen; (B) The fish lice, *Argulus*

other parts of the body. The head of the parasite is buried in the tissues of the fish to suck out the blood of the host. The majority of carps are subject to infection particularly during the summer months, but very heavy infection is rarely met with. When fry and fingerlings are infected, their growth slows down.

When a few fishes alone are affected, mechanical removal of the parasites by pulling them out from their anchorage by means of fine forceps would suffice, preferably followed by a week permanganate bath for two to three minutes. Repeated application of 'Gammexane' effectively eradicates the parasite.

Ergasilus is generally found attached to the gills in several of the fresh-water and estuarine fishes, but is not known to be very harmful.

Numerous internal parasites of fishes are known, but none of them causes large-scale mortality. Some flukes (flatworms) pass their larval stages in the body of the fishes which thus act as intermediate hosts.

Fishes are subject to several kinds of disease, not caused by parasites. Of these, the most important is the gas disease caused by excess of nitrogen or oxygen. Excess of the latter due to an abundance of vegetation in the water will sometimes result in accumulation of gas beneath the skin, on the fins, around the eyes, in the stomach and intestine, or in the choking of the blood capillaries. Both large and small fishes are affected, but if they are transferred in time to fresh water, they revive.

Large-scale mortality of fishes sometimes occurs in ponds usually in summer, or when there is an algal bloom in the water. Continuous sultry weather followed by sudden showers favours such mortality. Water stagnates completely without any circulation. The algae multiply, forming a dense bloom. The bottom layers are fast depleted of oxygen by quantities of decaying organic matter augmented by algae dying and settling down at the bottom. The respiratory demands during night markedly deplete dissolved oxygen in water and very early in the morning only the topmost layers of water contain some oxygen. When sudden rains fall under such circumstances, the surface and bottom waters get mixed and the dissolved oxygen content of the mixed water falls below the minimum required for the fish to live. All except air-breathing fishes soon come up gasping for breath, and die of asphyxiation.

This fatal depletion of oxygen usually takes place in the early mornings. If the distress is detected in time, the stock of fish may either by transferred to other ponds or the water in the pond aerated by beating up the surface water with poles. As the day advances, oxygen will be replenished by the algae, reviving the fishes.

Some algae may cause mortality among fishes by choking the gills or by producing toxic substances in the water.

Chapter 5
Broodfish Care

Care bestowed on the nurture of broodfish is the key to the success of induced spawning of the Exotic and Indian major carps. On it depends such vital processes as onset of maturity, full development of gonads and response to hypophysation. Also affected are the extent and rates of ovulation, fertilization and hatching, and the health of hatching and even their survival and growth. Under extreme case of improper broodfish care, gonads may not develop in the adult fish to any appreciable extent.

Sources Broodfish

A distinction was made between a stock pond and a broodfish pond. The functions of the two are different. A stock pond is devoted to growing and fattening of the fish to marketable size whereas, a broodfish pond is meant to hold fish in preparation for spawning. Maturing fish netted from a stock pond can also be a source of fish for the broodfish pond. Other sources of indigenous fish for the broodfish pond can be capture fishery waters such as rivers, lakes and reservoirs. In the case of exotic species (such as all Exotic carps in South and Southeast Asian countries), the only sources of broodfish, at least for the time being, are the stock ponds of fish farms and hatcheries.

Characteristics of Broodfish Ponds

A broodfish pond, depending on the magnitude of operations, may vary in area from 0.2 ha to 2.5 ha and, for the sake of ease of netting operations, should be of a rectangular shape, with width not exceeding 25m. Its water depth may vary from about 1.5m to about 2.5m. A broodfish pond must not only be drainable but should be such that it can be subjected to periodical flushing with good clean water and be amenable to treatments of manuring and fertilization of its bottom for generation of plankton, particularly zooplankton, for proper gonadal development. An alternate source of cool water (25°-27°C) which may be an underground source like an artesian well, should be available to flush the broodfish pond. This facility sometimes conditions broodfish better for induced breeding. While the advisable range of stocking of a broodfish pond is 1,00 to 3,00 kg per hectare, the lower rate of 1,00 kg per ha a preferable and should be used when space is available.

The number of broodfish ponds in a hatchery depends on methods of dealing with different species. The alternatives open are: (1) nurture of a mixture of broodfish of both sexes of all the species of Exotic and Indian major carps in the same pond; (2) segregation of a combination of two or more species of both sexes in the same pond; (3) segregation of both sexes of a single species and (4) sex-wise segregation of different species in separate ponds.

Fish species which breed naturally in a pond must be kept separately from those which do not since the former require segregation by sex in order to avoid uncontrolled spawning. Among the fishes under consideration in this book, the common carp, *Cyprinus carpio*, is the only naturally breeding fish under pond conditions; the rest need to be induced to breed in captivity. Further, the breeding season of common carp is often, at least partly, different from those of Exotic and Indian major carps. Since common carp breeders must be segregated sex-wise, separate ponds are required from females and males of the species. As to the rest of the species, any of the four above–stated alternatives can be adopted. Nature of different species of broodfish of Exotic and Indian major carps in one and the same pond is problematic since in a community broodfish pond, only common artificial feeds can be administered, whereas, the natural food and artificial feed requirements of the different species are separate for their respective proper gonadal development. Therefore, since the primary aim of broodfish ponds is to allow proper gonadal development of spawners, more especially the development of ovaries in the females, broodfish care is best done if the different fish are kept segregated, single species-wise and sex-wise. If the sexes of the brood Exotic and Indian major carps are individually segregated in separate broodfish ponds, the females would need much greater care at higher rates of feeding than the males.

In the unavoidable necessity of combining species owing to lack of space, the following combinations of species would be acceptable because of feeding complementarity; (1) broodfish of silver carp may be co-stocked with those of rohu or common carp but silver carp broodfish should not be stocked together with rohu and common carp in the same pond; (2) broodfish of grass carp may be co-stocked with those of catla or bighead carp but grass carp broodfish should not be stocked together with catla and bighead carp in the same pond; (3) bighead carp, catla and silver carp, which have overlapping feeding habits, should not be co-stocked in the same broodfish pond. If they are so stocked owing to overriding considerations of pond space, the zooplankton feeders (bighead carp and catla) should proportionately be much less than the silver carp. It is, however preferable to separate the broodfish of planktophagous species from each other and stock them in individual broodfish ponds according to species and sexes.

If pond space is extremely limited, it may become necessary to have one or more community stock ponds. Apart from saving space, such a community pond would allow fuller use of different ecological niches and enable specimens of different species to the procured for breeding operations through common seining operations. Should it become necessary to co-stock Exoitc and Indian major carps in one of common broodfish pond or ponds, their ratios may be approximately as shown in Table 5.1 with the overall stocking density remaining at 1,00 kg/ha to 3,000 kg/ha.

Adult catla often shows poor gonadal development. There is evidence that this can be caused by insufficient space. Catla brood fish need special care which would comprise; monospecies stocking at not more than 1,000/ha in the largest and deepest of the broodfish ponds: special steps like organic manuring; steps to ensure zooplankton production of a least 50 ml per 1,000 liters of water and feeding with a high protein diet (30 per cent) from which oilcakes may be excluded, over the greater part of the year. While scientific evidence is lacking, it is believed that presence of oilcake in the supplementary feed of catla tend to deposit fat in association with gonadal tissue in the particular zooplanktophagous fish. A catla brood pond also needs fortnightly flushing with clean freshwater to simulated riverine conditions.

Table 5.1: Ratios of Different Species of Carps in a Community Broodfish Ponds

Niche/Natural food	*Species*	*Percentage of Stocking*
Surface Feeders		
Phytoplankton	Silver Carp	24
Zooplankton	Bighead carp	12
Zooplankton	Catla	12
Bottom Feeders		
Detritus	Mrigal	12
Bottom-column Feeders		
Browser	Rohu	20
Macro-Vegetation Feeders		
Aquatic Foliage	Grass carp	20
	Total	100

The following management practices should be applied to broodfish ponds, whether stocked with one or more species.

Pond Fertilization and Manuring

The objective of pond fertilization and manuring is to produce a plentiful, sustained supply of plankton and, in the context of broodfish care, more of zooplankton to provide the animal protein required in the diet of the broodfish. The cycle of phytoplankton and zooplankton production is linked with pond productivity and growth of fish. Organic manures such as cattle dung, chicken and duck droppings and composted water hyacinth with urea and lime produce zooplankton biomass. It is a sound practice to periodically assess plankton production and to take corrective measures through manuring should the level of production run low. Secchi disc transparency, in otherwise non-turbid waters, can be an added tool to judge plankton production. The entire pond bottom treatment is based on soil chemistry and should be so treated in a hatchery. The objective should be to supply only the missing nutrients in a particular agroclimatic conditions.

Fish Feeds and Feeding

Sources of protein, carbohydrates, vitamins and minerals and somewhat variable and are, up to a point, dependent on local availability. For example, soyabean, as a source of protein, is not easily available in all countries; silkworm pupae are available in only a few countries, and only in certain areas within these countries; fishmeal is also not a universally available commodity and its price is highly variable. A possible broodfish supplemental feed includes soyabean, meal, corn and rice bran in the ratio of 60:10:80, respectively, with a total protein content of about 30 per cent. Ingredients and percentage composition of supplemental feed for common carp broodfish are shown in Chapter 8.

The feeding rate may be 1 per cent to 2.5 per cent of the body weight of broodfish during the autumn and winter months of September of January depending on water surface available for each broodfish to graze nutrient food. Later, it would become necessary to raise the feeding rate (feed remaining the same) to 5 per cent of the bodyweight of the fish. Feed should be administered once a day, every day of the week and at a fixed time and fixed location. It may be handmade or palletized and may be had-broadcast or served through feeders. Some aspects of feeds for broodfish are mentioned in Chapter 8.

There is a practice followed in China for conditioning broodfish of Chinese carps aimed at minimizing broodfish mortality in induced breeding operations. This practice comprises drag-netting broodfish into a corner of the broodfish pond and putting them in a net enclosure for about one hour and letting them go. This process is repeated for two to three days before the day when inducted breeding made of canvas from the broodfish pond to the hatchery proper.

Special Dietary Care of Broodfish

Special dietary care is required for the different species of broodfish. Guidelines are given below, as far as known.

Grass Carp

The spawners of grass carp are to be fed once daily, during autumn and spring, on wheat and paddy sprouts, corn grains and bean cakes in equal proportions at 1-2 per cent of their body weight besides macrovegatation at 100 per cent of their body weight. The first-mentioned feeds are believed to enhance fecundity of grass carp and promote proper gonad development.

Bighead and Silver Carp

For these planktophagous fish fertilization with organic manures at the rate of 1.5 to 2 tonnes per hectare every ten days in recommended. Should the gonad development remain poor, supplementary feeding with powdered bean cakes, peanut cakes and wheat or rice bran in equal proportion at 1-2 per cent of body weight per day may be given.

Catla

For proper gonadal development of catla (besides the earlier mentioned precaution of monospecies stocking and flushing of broodfish ponds), organic manuring with cattle dung and/or chicken and duck droppings, and production of zooplankton at 30-5–ml per 1,000 liters of water are very important. Artificial feeds containing soybean meal and/or fish meal (30 per cent) with rice bran or wheat bran but without oilcakes administered at 3-5 per cent of body weight daily are important.

Mrigal, Rohu and Common Carp

These species are by and large bottom dwellers and are naturally detritus eaters or browsers. Feeds based on soybean meal, rice bran or wheat bran and oilcakes in equal proportions at 3-5 per cent of body weight per day benefit their gonadal development. An alternate feed at 3-5 per cent of body weight per day is a mixture of wheat bran, mustard oilcake, coarse wheat flour and fish meal or soybean meal in the ratios of 4:4:1:1 Mustard oilcake, if used, should be presoaked (for softening) in water overnight.

Chapter 6

Induced Spawning of Indian Major and Chinese Carps

The Indian major carps (IMC) and the Exotic Chinese carps breed naturally only in the flowing waters of their natural habitat, the rivers, in association with the mansoon floods. Until the introduction of the technique of induced breeding, the only source of their seeds for aquaculture was the rivers. The riverine crap spawn invariably is a mixture of seed of desirable and undesirable species of fish and its availability entirely depends on the vagaries of the monsoon. On the other hand, induced breeding produces seed of absolute purity and, at the present stage of the technical development of induced breeding, allows at least partial independence from the monsoon. It also opens the possibilities of stock improvement by selective breeding and hybridization. The progress of carp culture in recent years is almost entirely due to the development of induced breeding techniques.

Stages of Maturation of Adult Carps

Seven stages of gonadal maturation of carps are commonly recognized. The stage and their characterizes are shown in Table 6.1.

After the third state maturation has been reached, further development of ovary is very fast, especially in increasing day-length and rising temperature. By the time Stage V is reached, the ovarian weight may be 20-30 per cent of the total body weight (*i.e.*, gonado-somatic index : 20-30). In males, the gonadosomatic index may be 5-10 at stage V of maturation. Common carp exhibits a more pronounced ovarian development than Chinese or Indian major carps with females showing a gonado-smatic index of about 30 and males, 20-30.

Table 6.1: Stages of Gonadal Maturation of Carps

Sl.No.	*Stage of Maturation*	*Condition of Gonad*	
		Female	*Male*
I.	Immature	A transparent, long, narrow strip	Transparent, thread-like
II.	Immature	Thickened strip, translucent	Translucent, thread-like
III.	Maturing	Opaque, granular, some what grayish, occupying about 1/3 of the body cavity	Opaque, pinkish white, thin, strip-like
IV.	Maturing	Dull grayish in colour, ova discernible like granules, occupying about ½ of the body cavity	Thick strip, milky, oozing whitish fluid on applying pressure on belly
V.	Mature	Dully grey to greenish, occupying almost the entire body cavity, ova distinctly round in shape	Thickened band like oozing whitish fluid on applying pressure on belly
VI.	Spawning	Having loose eggs in ovarian wall, eggs oozing through genital aperture on applying pressure on belly	Oozing milt freely on applying
VII.	Spent	Ovary blood-sort, pinkish brown mass but greatly shrunk	Oozing milt freely on applying pressure on belly

While sexes in these carps are morphologically different, the sexual dimorphism they exhibit externally is only relative and not infallibly distinct even in the breeding season. The more important external distinguishing features of ripe female and male Chinese carps and Indian major carps are shown in Table 6.2.

Table 6.2: External Distinguishing Features of Sexes of Carps

Sl.No.	*Female*	*Male*
1.	Pectoral fin relatively small and week with the outermost ray not very thick	Pectoral fin relatively long and prominent with well-developed thick outermost ray.
2.	Inner surface of the pectoral fin facing the body, is smooth to feel by touch.	Inner surface of the pectoral fin, facing the body, is rough to touch.
3.	Abdomen shows a conspicuous bulge which extends past the pelvis up to genital aperture. Abdomen soft to touch. No median ridge in front of vent. Bulge in the abdomen may also de due to fat deposit around the gut and is not a fool proof indication of mature gonads.	Abdomen does not generally show a conspicuous bulge and is not very soft to touch. Abdomen shows median ridge of vent.
4.	Genital aperture is protruding and swollen, turgid, shows pinkish margins. A pinkish genital pore may also be found among fish which have begun to regress. (This method is particularly difficult to apply to grass carp)	Vent not protruding; pit-like in appearance.
5.	Ova visible inside genital aperture when gentle pressure applied to abdomen. Vent may also be swollen and reddish.	Milky white milt exudes through genital aperture applying gentle pressure to abdomen.
6.	Body stouter in appearance relative to male of same age adult.	Body thinner and linear in shape relative to adult females of same age.

Determining Readiness for Spawning

The morphological features described in Table 6.2 can be taken as indicators of ripeness. However, a more reliable method or determining whether or not the female fish a ready for spawning is to draw an egg sample from the posterior end of the ovary by means of a catheter. A catheter, for this purpose, is a glass or plastic tube of 2 mm diameter with a short flexible tubing attached to one end to permit easy movement. The catheter is inserted into the genital aperture and probed to reach the posterior extremity of the ovary.

An egg sample is then sucked into the glass tube and transferred to a glass bowl into which a solution of 70 per cent acetic acid and 30 per cent alcohol is immediately poured, immersing the ovarian ova sample. Within about 5 minutes, yolk becomes discernible and so also the nuclei of individual ova. Acentric or peripheral location of nuclei is a sure indication of readiness of the fish for spawning. In gametogenesis, the nucleus of the ovum migrates to the periphery before release of the first polar body in meiosis. Should, however, the ova nuclei be found to be centrally located, then such a fish would not respond to hormonal breeding treatment.

Induced Spawning of Chinese and Indian Major Carps

The most commonly adopted technique of induced breeding of carps is hypophysation. It involves injecting mature female and male carps with extract of pituitary gland taken from other mature fish generally phylogenetically close to carps. The pituitary gland (hypophysis) is an endocrine gland located on the ventral side in many case, the pituitary gland is lodged in a distinct depression on the inner side of the cranium of the fish.

In order to prepare fresh extract of pituitary, the gland has to be removed by dissecting the fish. The best way to get at the whole

Figure 6.1: Genital Aperture (Protruding and Swollen) of a Mature Female Rohu

gland is to saw or chop off the roof the cranium horizontally, exposing the brain. The brain is then carefully removed from the cranial cavity with a pair of forceps. This process exposes the pituitary gland (located ventral to brain) from the dorsal side. The gland is then picked up in whole with a pair of forceps (Figure 6.2). Care has to be exercised to ensure that the gland is not damaged in the process of removal and collection, since otherwise some of the water-soluble gonadotropins may be lost.

If the gland is not to be used immediately for preparation of an extract for fish breeding, it may be preserved. The preservation of whole gland is a important step to ensure retention of their potency. Only entire gland should be preserved. The tissues that may be attached to the gland should be removed and the glands either preserved in absolute alcohol or acetone-dried.

If pituitary glands are preserved in absolute alcohol, airtight vials should be used and after being kept in this medium for 24 hours, old alcohol should be replaced by new absolute alcohol to ensure proper preservation. The vials then may be stored even at room temperature to retain their potency up to a year but the use of a

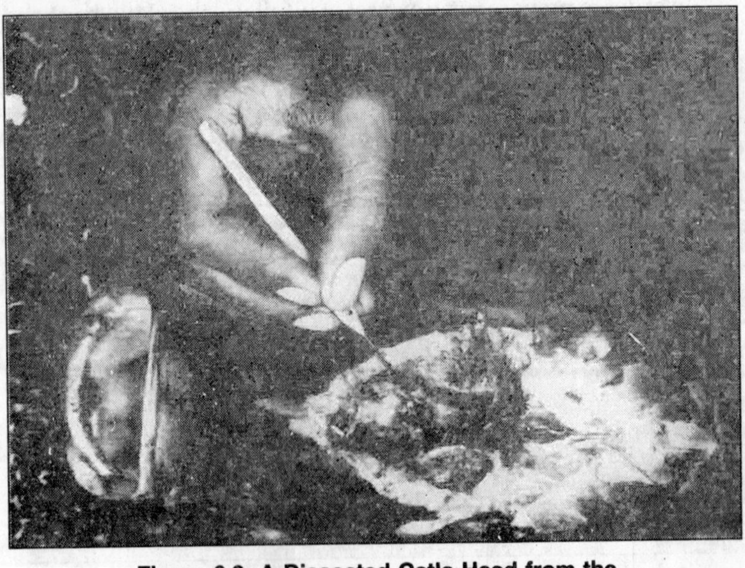

Figure 6.2: A Dissected Catla Head from the Dorsal Side to Show the Pituitary Gland

refrigerator is preferable is it allows the preserved glands to be retained for two to three years with impairing their efficacy.

If acetone-drying is to be done, the glands should be placed in 10 times their volume of acetone, and the latter should be replaced with new acetone after 12 hours. The acetone should be changed again after a further 6-8 hours. After this, the glands are taken out, dried on filter paper and stored in tightly sealed vials for future use. Pituitary glands preserved in this way are known to retain their efficacy for over three years.

There are certain precautions to be taken in choosing the donor fish for collection of pituitary glands. It is best to choose freshly killed fish. Next best is fish freshly preserved in ice or cold storage but preserved well before any spoilage has started. Pituitary glands should be collected only from adult, sexually mature unspent fish. Gonadotropin from pituitary extracts of phylogenetically remote fish such as salmon, tilapia, catfish and even some marine fish has been found to be effective in carps; however, common carp pituitary and homoplastic pituitary are perhaps safest to use.

Preparation of the Injection Material

In early year of induced breeding, the weights of donor and recipient fish were kept approximately equal. This meant that for spawning a 2 kg female carp, for example, the pituitary gland extract from a donor fish also weighting about 2 kg was necessary. As a result of further research in fish breeding, the required weight of pituitary gland extracted from a donor fish can be calculated based on the total weight of the recipient fish (Figure 6.3), so that the weight of the donor fish need not be known. A system of two injections has been evolved in which the first injection called a stimulating or priming dose, consists of a preliminary of 2 to 4 mg of pituitary per kg of the body weight of the recipient female, following six hours later, by a second, called a resolving, of 8-12 mg/kg body weight. In this manner, a total of 10-16 mg/ kg body weight is injected to the female spawner. For recipient male fish, a single dose of 2-4 mg/kg body weight is given at the time of the second injection of the female fish. The system of two injection to recipient fish has, by and large remained unchanged since the early sixties when it was first developed in India. The injection dosages may have to be varied depending on the exact stage of maturity of the recipient fish and

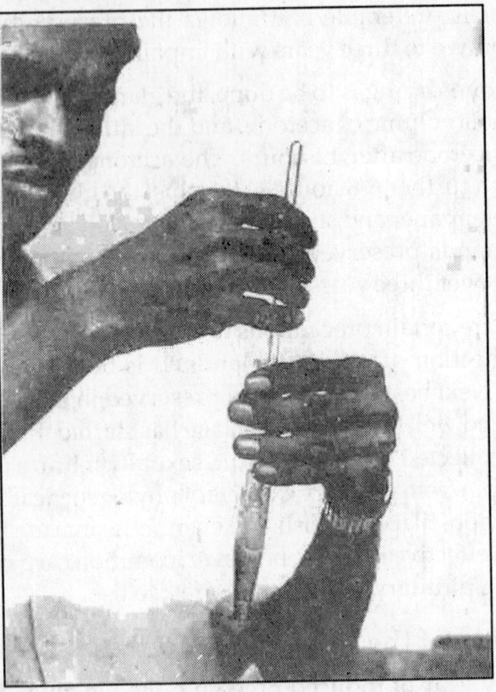

Figure 6.3: A Recipient Catla Female being Weighed to Determine the Dose of Pituitary Extract to Injected

water temperature, a relatively lower stage of maturity and cooler water required greater dosage of injections. The extent of variation in dosage is a matter of conjecture in the present day state of technical knowledge of hypophysation of carps.

For injection, the glands should be gently removed from the vial with a pair of forceps, dried on filter paper for about 2-3 minutes if preserved in absolute alcohol, and then weighed on an electric balance. The weighed glands should be transferred to a tissue homogenizer (Figure 6.4) and thoroughly macerated in a few drops of distilled water. More distilled water should then be added to give about 40 mg of pituitary material per ml of water. The extract is then centrifuged and the supernatant fluid is used for injections. Requisite amounts of fluid according to the weight of the recipient fish may be drawn into a syringe for injections.

Figure 6.4: Carp Pituitary Gland being Homogenized

Injection Routes

Injections may either be intramuscular (Figure 6.5) or intraperitoneal. Given equal degree of dexterity of the worker with both, the quantity of fluid to be injected indicates which of the two routes should be used. The intramuscular route is effective when the fluid to be injected does not exceed 2 ml. For greater volume, the intraperitoneal route is more appropriate because the peritoneal cavity has much more space in which to hold the injected fluid than does muscular tissue.

Intramuscular injection are generally given between the base of the dorsal fin and the lateral line, at the base of pectoral fin or between the anal fin and lateral line. The fish scale should be lifted for inserting the injection needle, and the area should be gently massaged following the withdrawal of the needle after injection to aid distribution of the extract into musculature and prevent its backflow.

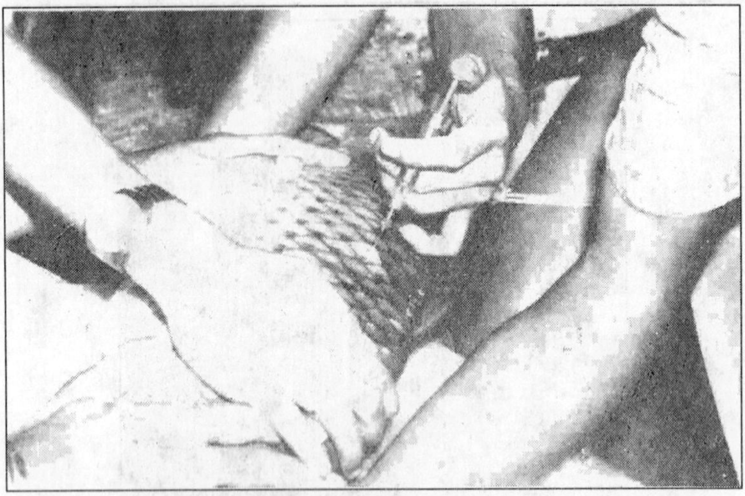

Figure 6.5: Intramuscular Injection being given to a Catla Broodfish

Backflow can also be reduced by increasing the viscocity of the fluid; this can be done by adding glycerine in the ratio of 7:3 (water:glycerine). The placement of the needle for the intramuscular injection is unimportant for Indian major carps, but its angle to the base of pectoral fin is recommended as 45º for silver and bighead carps and 90º C for grass carp. Intraperitoneal injection may be at the base of either the pectoral fin or the pelvic fin. If it is the former, care has to be exercised that the heart of the fish is not punctured by the needle. It is preferable to use a smaller needle for intraperitoneal injection at the base of the pectoral fin. The size of the syringe and the diameter (gauge) of the needle depend partly on the size of the recipient fish but more so on the volume of the fluid to be injected. Gentle handling of the broodfish during stage in preparation for injection or while injecting is of great importance since bruising, abrasion and scale-loss, or even removal of mucous, render skin liable to bacterial infection. A careless handled fish may not survive long enough to ovulate, whereas a fish that is handled fish may be successfully bread several times. A moist towel can be used to wrap the fish for transport even over short distances. A 0.01 per cent acriflavin dip after handling fish for injection or stripping reduces problems of bacterial infections.

Use of anaesthetics during injection would significantly increase the survival of broodfish. Commonly used anaesthetics are MS222 (Sandoz) and Quinaldine. MS222 may be added to the water in doses of 50-100 mg/liter or a roll of cotton-filled cloth may be soaked in a 0.04 M solution and inserted in the mouth of the fish. If Quinaldine is to be used as a anaesthetic, its dose is 50-100 mg/l. The injection can also be given while the fish is in water, a process which eliminates the risk of internal injury since the weight of the fish is borne by the water; this may be done either on anesthetized fish or unanesthetized fish held in the folds of a net or towel in a hinged padded fish holder.

In fish farms and hatcheries where artificially controlled environmental conditions are not available, hypophysation work should be done on cloudy or rainy days. The first injection is generally given around 1600 to 1800 hours and the second, six hours later. Such timings generally suit the convenience of workers and the diurnal temperatures are also the lowest when spawning condition of the injected fish is expected (0500 to 0700). These consideration of timing and weather are of vital importance, especially where breeding and hatching tanks are located in the open, as otherwise the mid-morning sun could cause the water of cement breeding and hatching tanks to heat up and destroy the ovulated eggs totally. In locations at higher altitudes where the ambient air and water temperature and lower than those on the plains, with cold nights intervening between successive sunny days, sunny days may be chosen for breeding. In such cases, however, soon after oviposition in breeding tanks, the eggs should be transferred to the hatching tank and the flow of water must continue in both, to avoid mortality of the eggs and hatchlings caused by heat. In cases of indoor hatcheries where controlled environmental conditions exist, the above injection schedule need not be followed.

Injected broodfish of Exotic and Indian major carps and released in breeding tanks in appropriately paired sets. Generally, a set is made up of one female fish plus two males when combined weight approximately equals the weight of the female. In the flowing water of a cemented breeding tank, whatever its shape, spawning and fertilization generally take place naturally within six hours after the second injection of pituitary extract. If six hours have elapsed since the second injection, and spawning has not taken place naturally in spite of the provision of optimal conditions, artificial insemination or stripping has to be used. Many hatcheries have developed specific injection and treatment procedures for different species of fish, especially the Chinese carps, which necessitate stripping at a specified time after the injection are administered. These procedures are described later in this book for each species.

For a female broodfish to be ready for stripping, the hormonal injections must have induced ovulation of the ova within the ovary. This is indicated by a very soft belly condition at any part of the abdomen, and a state of shifting softness should be felt when the worker's hand is moved along the abdomen. Internally, the individual ova must have been discharged from their follicles inside the ovary wall. In this condition when the abdomen is pressed by hand, a jet or spurt of loosened eggs would occur through the genital aperture. For stripping, the female fish is held by the operator between the side of the body and his arm with the fish slanting head up, tail down and belly facing the vessel (Figure 6.5) and the eggs are collected into an enamel or plastic trough by pressing the body of the fish. The male fish is then similarly held and milt is squeezed out

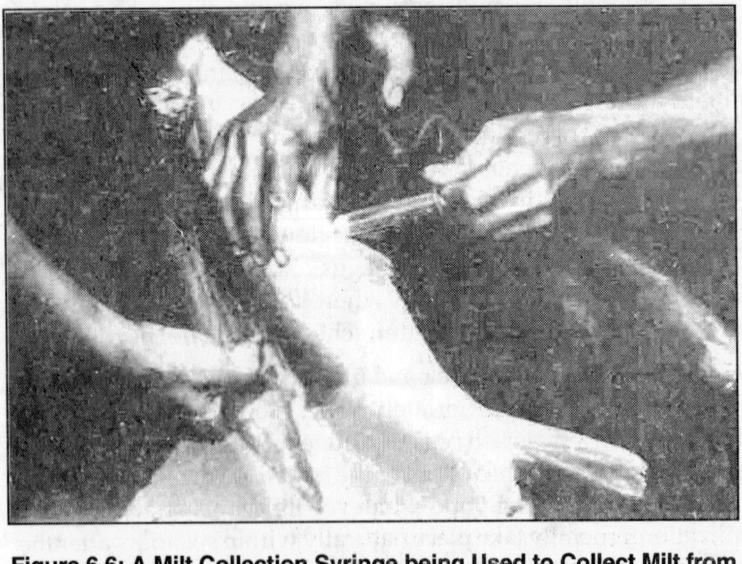

Figure 6.6: A Milt Collection Syringe being Used to Collect Milt from an Adult Male Carp (Reproduced from a filmstrip entitled "Rearing fry and fingerlings of Chineses carps" FAO)

into the same basin as the eggs. Two persons can more easily carry out stripping, with one holding the fish firmly and the other liberating the eggs and milt by turns. At least three persons are required for simultaneous release of eggs and milt. The two sex products are then mixed as soon as possible by means of a feather to allow fertilization to take place. The fertilized eggs are then washed a few times with clean water to remove excess milt and allowed to stay undisturbed in freshwater for about 30 minutes. The eggs are then ready for release in the hatching tanks (Figure 6.7).

Figure 6.7: Stripping of a Female and a Male Chinese Carp

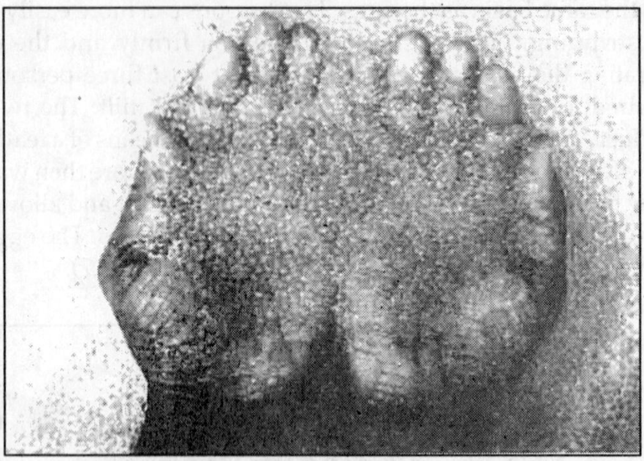

Figure 6.8: Fertilized Chineses Carp Eggs

Other Ovulating Substance for Induced Spawning of Carps

The technique of hypophysation has undergone refinements over the years beginning with preliminary experiments performed by Houssay in 1931. At present, hypophysation remains the only realistic choice for the rural fish farmer of Asia, for reasons both economic and technical. It is however, beginning to appear that the days of hypophysation for carp breeding work are numbered. The problems of hypophysation are mainly those of dosage and the supply of pituitary-based gonadotropic hormones; the former arises out of the crudeness of the technique and the latter, out of supply related to demand (apart from the time and money required for collecting large amounts of pituitary glands from maturing adults of phylogenetically close fish in a short season.) Significant development in this context are the dehydration and ampouling of crude pituitary extract of homoplastic material (Figure 6.9) attempted establishment of pituitary banks by the Food and Agriculture Organization of United Nations with assistance from the Indian Council of Agricultural Research (ICAR) and the availability of powdered crude extract of common carp pituitary from some pharmaceutical companies. These developments do not solve the problem of standardization of dosages for which accurate knowledge

Figure 6.9: Ampoules Containing Fish Pituitary Extract Preserved in Glycerine

of the gonadotropic activity of pituitary extracts and development of suitable bioassay techniques are essential.

Various substance are at the present time used in fish breeding work. In Thailand, pituitary glands from *Tachysurus* sp. and *Trichogaster pectoralis* are also used. Human chorionic gonadotropic hormone (HCG) is also widely employed (Table 6.3) although its excessive use can produce immunological effects.

Silver carp and bighead carps are bred in Malaysia at 25-27ºC water temperature by injecting a priming dose of 50 international units (IU) of HCG per kg followed by a stimulating doses of 200/250 IU HCG/kg 12-24 hours later, further followed 6 hours later by a resolving does of 4 mg/kg of acetone-dried common carp pituitary extract with ovulation occurring 6 hours later. In China, high doses of HCG are used at times for breeding bighead carp, silver carp and grass carp. The doses used are of the order of 500 to 2,200 IU/kg for bighead carp and 800 to 900 IU/kg for silver carp with grass carp needing even higher doses than these.

To circumvent the un-bioassayed character of the extract of whole pituitary glands, the weights of donor and recipient fish a equated in Hong Kong in the ratios of 0.9:1:4, 1.0:1:3 and 1.0 to 1.6 for bighead carp, silver carp and grass carp, respectively, and in Thailand, in the ratios of 0.4:1.0, 0.3:0.7 and 0.5 to 0.7 for the same

Table 6.3: Induced Breeding Techniques of Chinese Carps Adopted by Various Workers[a]

Species (Country)	Time of Year	Age of Breeder (Yrs.)	No. of Females	Water Temp. (°C)	Assessment of Gonadal State	Substance (Route)	Solvent	Injection 1st Done	Injection 2nd Done	ΔT[b] (hr)	Delay to Ovulation (hr)	Fertilization Method	Incubation Method	Results	Notes
Hypophthalmichthys molitrix (Israel)	–	2-4	60	–	F: enlarged soft belly M: milt on gentle pressure	*Cyprinus carpio* pituitaries (IM)	0.75% NaCl	–	–	–	–	Stripped: dry	Incubators with flowing water	3 million larvae hatched, 50% survival to 0.25 g stocking size	Injection carried out while fish in water to eliminate trauma
Hypophthalmichthys molitrix (Malaysia)	–	2	12	17-28	F: full, soft belly, swollen, pink cloaca M: milt on pressure	M or F *Cyprinus carpio* or *Puntius gonlonotus* pituitaries (IM or IP)[c]	0.6% NaCl	F:5-6 mg/kg N : 5 mg/kg	F:5-6 mg/kg N:5 mg/kg	5	5-6	Stripped : dry	In hatchery trays with running water; in ponds: stagnant hapas	Hatchlings in excellent condition for 30% of females injected	Addition of Synahorin to extract has little effect
Hypophthalmichthys molitrix (Malaysia)	–	–	–	25-27	–	HCG plus *Cyprinus carpio* pituitaries	–	F:200 IU HCG/kg	F:4 mg/kg *Cyprinus carpio* pituitary	6	6-7	–	–	80% successful ovulation	
Hypophthalmichthys molitrix (China)	–	–	198	20-28	–	LHRH-A[d]	Water base	F: 8-10 mg/kg M:5 mg/kg	–	–	22-24	–	–	84% induced ovulation	

Contd...

Table 6.3–Contd...

Species (Country)	Time of Year	Age of Breeder (Yrs.)	No. of Females	Water Temp. (°C)	Assessment of Gonadal State	Substance (Route)	Solvent	Injection 1st Done	Injection 2nd Done	ΔT^b (hr)	Delay to Ovulation (hr)	Fertilization Method	Incubation Method	Results	Notes
Hypophthalmichthys molitrix (India)	Jul-Aug	–	40	28-34	–	Homoplastic and heteroplastic pituitaries (IM)	–	F: 4 mg/kg	F:6-10 mg/kg M: 2-3 mg/kg		4-5	Stripped Stagnant hapas	Stagnant hapas	5-80% fertilization : 907,000 fry obtained	Addition of 25-50 IU Synahorin to both female injections also effective. Maximum fry yields associated with rain and cool weather, Best water temp. 27-30° C
Hypophthalmichthys molitrix (Thailand)	May-Aug	1-2	10	27-33	–	*Cyprinus carpio* pituitaries plus Synahorin (IM)	0.6% NaCl or distilled water	F: and M: 0.23-1 pituitary	F:1-34 pituitaries plus 20 rebbit units Synahorin	6-8	4-6	Stripped: dry	Hapas in running water	67.7% ovulation	
Ctenopharyngodon idella (Malaysia)															

Contd...

Table 6.3–Contd...

Species (Country)	Time of Year	Age of Breeder (Yrs.)	No. of Females	Water Temp. (°C)	Assessment of Gonadal State	Substance (Route)	Solvent	Injection 1st Done	Injection 2nd Done	ΔT[b] (hr)	Delay to Ovulation (hr)	Fertilization Method	Incubation Method	Results	Notes
Ctenopharyngodon idella (China)	–	–	25	20-28	–	LHRH-A	Water base	F: 5-10 mg/kg M: 2.5-5 mg/kg	–	–	15-20	–	–	88% spawning	Total 139 fish treated with LHRH-A under various injection regimens, with a resulting overall spawning race of 86.3%
Ctenopharyngodon idella (India)	Jul-Aug	–	11	28-31	–	Homoplastic and heteroplastic pituitaries (IM)	–	F: 3-5 mg/kg	F: 7-10 mg/kg M: 2-3 mg/kg	–	5-6	Stripped : dry	Stagnant hapas	15-80% fertilization; 553,000 fry produced	Combining Synahorin (25 IU) with pituitary extract proved ineffective
Ctenopharyngodon idella (Thailand)	Jul-Oct	1-2	8	27-33	–	*Cyprinus carpio* pituitaries plus Synahorin (IM)	0.6% NaCl or distilled water	F and M 0.23-1 pituitary	F: 1-3.4 pituitaries plus 20 rabbit units Synohorin	6-8	4-6	Stripped: dry	Hapas in running water	47% ovulation	

Contd...

Table 6.3–Contd...

Species (Country)	Time of Year	Age of Breeder (Yrs.)	No. of Females	Water Temp. (°C)	Assessment of Gonadal State	Substance (Route)	Solvent	Injection 1st Done	Injection 2nd Done	ΔT[b] (hr)	Delay to Ovulation (hr)	Fertilization Method	Incubation Method	Results	Notes
Aristichthys nobilis (Malaysia)	–	2	18	27-28	F: full, soft belly, swollen, pink cloaca M: milt on pressure	M or F *Cyprinus carpio* or *Puntius gonionotus* pituitaries (IM or IP)	0.6% NaCl	F: 5-6 mg/kg Wet weight	F: 5-6 mg/kg M: 5 mg/kg	5	5-6	Stripped: dry	In hatchery: trays with running water; in ponds: stagnant hapas	Hatchlings in good condition for 66% of females in injected	Addition of Synahorin to pituitary extract has little effect
Aristichthys nobilis (Malaysia)	–	–	–	25-27	–	HCG plus *Cyprinus carpio* pituitaries	–	F: 200 IU HCG/kg	F: 4 mg/kg *Cyprinus carpio* pituitary	6	6-7	–	–	80% successful ovulation	–
Aristichthys nobilis (China)	–	–	108	20-28	–	LHRH-A	Water base	F:1-2 mg/kg M: 0.5-1 mg/kg	F: 8-9 mg/kg M: 4-4.5 mg/kg	12	–	–	–	82% induced ovulation	
Aristichthys nobilis (Thailand)	Aug-Sep	1-2	5	27-33	–	*Cyprinus carpio* pituitaries plus Synohorin	0.6% NaCl or distilled water	F and M: 0.23-1 pituitary	F: 1-3.4 pituitaries plus 20 rabbit units Synahorin	6-8	4-6	Stripped: dry	Hapas in running water	91% ovulation	

a: Table based on Harvey and Hoar (1979); b: T = Time between first and second injections; c: IM = intramuscular; IP – intraperitoneal; d: LHRH-A = (D.Ala[6].Des-Gly-NH_2[10]) – LH-RH-ethylamide.

fish, respectively. The amounts of extracts required for induced spawning of phylogenetically distant species are greater than for more closely related species; for example in Malaysia, bioassayed salmon gonadotropin was found to be much less effective on Chinese carps than common carp pituitary extract.

Research is currently underway on a number of these substances, such as synthetic luteinizing hormone-releasing hormone analogues (LH-RHA), clomiphene citrate, tamoxiphen, progesterone, DOC (deoxycorticosterone), which show some promise as ovulatory agents. At the present time, it appears too early to give firm recommendations regarding these substances.

Based on knowledge gained from different countries, preferred hormonal treatments to the female spawners of different species of Chinese and Indian major carps may now be recommended. The males in every case to be injected at the time of the second injection to the female.

Protocols for Pituitary Injections to Induce Spawning in Indian Major Carps (IMC)

Rohu (*Labeo rohiṭa*)

A priming dose of 2 mg/kg of broodfish at 1600 to 1800 hours followed by a resolving dose of 8 mg/kg, 6 hours later. Natural spawning may be expected 4 to 6 hours after the second injection. Optimal water temperature is 27°C, within the acceptable range 24°C to 31° C. Rohu is not intolerant to warmer water within its acceptable temperature range.

Catla (*Catla catla*)

A priming dose of 1-2 mg/kg at 1600 to 1800 hours followed by a resolving dose of 6-8 mg/kg, 6 hours later. Natural spawning may be expected 4 to 6 hours after the second injection. Catla is very intolerant to warm water and should be induced to breed during heavy rains when cool weather prevails. In the event of warm water spell, catla especially should be exposed to cool hatchery water of 27°C to 28°C for an additional day in the ante or storage tank before breeding by hypophysation.

Mrigal (*Cirrihinus mrigala*)

A priming dose of 1-1.5 mg/kg at 1600 to 1800 hours following by a resolving dose of 6 mg/kg 6 hours later. Natural breeding may be expected 4 to 6 hours after second injection. Optimum temperature required by mrigal is 27°C within the range of 24°C to 31°C. Mrigal is not intolerant to warmer water temperature within the acceptable temperature range.

In Indian major carps, if it is necessary to strip, because natural breeding did not take place on time, or if it is desired to hand-strip to obtain fuller oviposition, the injection venue may be changed to the ante-tank, administering both the injection at the same dosages as in the case of natural breeding. The fish should be examined for the freely oozing condition and softness of belly and stripping done between 4 to 7 hours after administering the resolving dose.

Protocols for Induced Spawning of Chinese Carps

A wide variety of substance and procedures of their use have been adopted by different workers for successful breeding of silver carp, bighead carp and grass carp. The various protocols tried in respect of different species of Chinese carps are shown in detail in Table 6.3.

Common Carp Breeding

Uncontrolled Breeding of Common Carp (*Cyprinus carpio*)

Common carp differs from Chinese and Indian major carps in its fecundity to breed naturally in ponds provided that some aquatic vegetation is available for attaching its adhesive eggs and that environmental condition are favourable. The fish breed throughout the year in tropical climates, (*e.g.*, in Indonesia) with two peak breeding period, one during spring often lasting from January to April and the other during autumn often lasting from July to October. Where the fish does not breed throughout the year, it generally breeds twice a year during spring and autumn. In Europe, it breeds during the spring and summer months, March to July.

For obvious reasons, the faculty to breed naturally should ordinarily be of great advantage for a cultivated fish. However, there are drawback in leaving fish to find for themselves and allowing them to breed in a uncontrolled manner, particularly high mortality among eggs, hatchlings and fry due to sub-optimal conditions (unsuitable temperature, high level of predation, etc.). To obviate these drawbacks of uncontrolled reproduction, common carp is bred in aquaculture in either a semi controlled manner as a generally the case, or in a controlled manner by hypophysation.

Semi-Controlled Breeding of Common Carp

There exist two major semi-controlled breeding systems for common carp in different countries of the world, and one can choose the one most relevant to a given situation. The two systems are the Indonesian and the European. Those followed in other countries fail in one of the other system or are at best minor alterations thereof.

The best results in common carp breeding are obtained when broodfish are carefully chosen. The following criteria for mature fish may be adopted for the "big belly" variety of common carps.

1. A fully mature female has an almost rounded, soft, bulging abdomen, with obscured ventral ridge and vent projecting into a small papilla-like out-growth.
2. A mature female, if kept on its belly, will rest thereon with out falling sideways, and when held with abdomen directed upwards, shows a slight sagging on the sides due to the weight of the developed ovaries.
3. Mature males (as in other cases) exude milt when gently pressed on the abdomen.

Broodfish are fed with rice bran, kitchen refuse corn etc. Spawning ponds are elongated, 25-30 m^2 in area and have a hard bottom devoid of mud and silt. The pond is dried for a few days before filling with clean water up to a depth of 50-70 cm. Water is released on the morning of the breeding date and broodfish as well as spawn collectors called kakabans are positioned in the afternoon. The Indonesian kakabans are made of dark horse-hair-like fibers of the Indjuk plant (*Arẹnga pinnata* and *Arenga saccharifera*). For making kakabans, the indjuk fibers are washed clean then layer thereof arranged in 1.20 to 1.50 m long strips (Figure 6.10). The long strips are joined lengthwise (Figure 6.11) between two bamboo planks 4-5 cm wide, 1.5 to 2.0 m long a nailed together on two sides. For spawning kakabans are kept in a floating position a little under the water surface, propped up on bamboo poles.

Five to 8 pieces of kakabans of the above description are required per kilogram weight of female spawners. A gentle flow of water is supplied in the breeding pond after the broodfish are released and

Figure 6.10: Making Kakabans from Split Cane and Mountain Moss

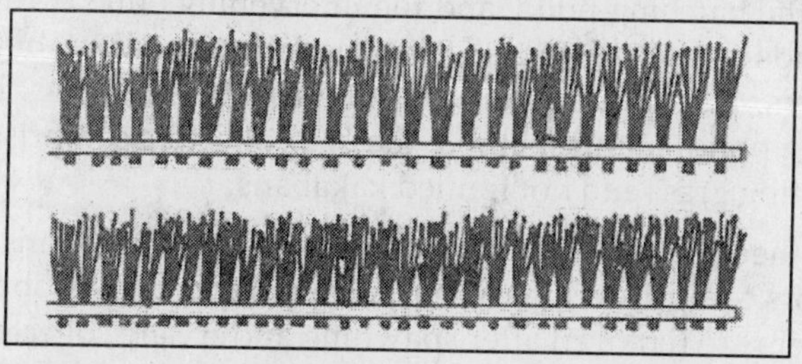

Figure 6.11: Indjuk Plant Fibers Arranged on Bamboo Strips

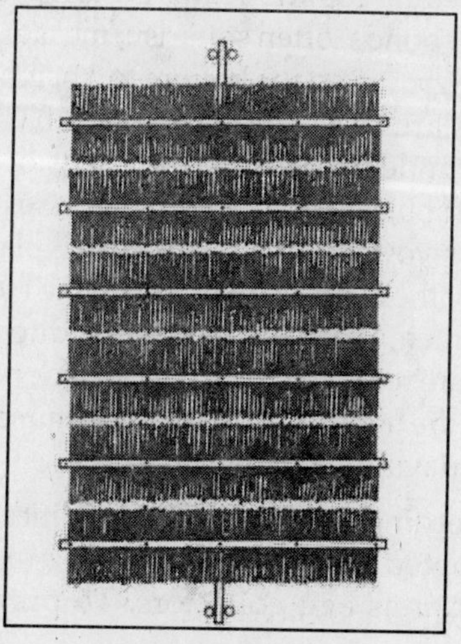

Figure 6.12: Kakaban Ready for Installation in Floating Ready for Installation in Floating Position

the kakabans installed. By habit, the fish first attaches its eggs on the underside of the kakabans. When the entire underside is fully of deposited eggs, the kakabans raft is turned over. When both sides of the kakabans are fully egg-laden, they are transferred to hatching ponds which are 20 times larger than the breeding pond. Here, the kakabans are placed transversely on floating bamboo poles leaving a gap of 5-8 cm between the fibers of the adjacent kakabans. Care is taken that the eggs always remain fully submerged in 8 cm of water. When three weeks old in the hatching pond, carp fry and collected by gradually lowering the water level. The expected yield of fry is 15,000 to 20,000 per kg weight of female spawners.

In the Rantjapaku method, which is also a variant of the Sundanese system, the bottom of the spawning pond is sandy and gravely. The spawning pond itself is located at a higher level than that of the hatching pond, and the intervening bund between the two ponds is made of stones so arranged that the spawn, but not the spawners, pass throughout the spaces between stones into the hatching pond. In this method, the egg collectors comprise newly cut floating grass and not framed kakabans.

In the Central Sumatra method, instead of kakabans, spawn collectors comprise scattered indjuk fibers spread over about 5 m^2. Spawners are removed after spawning and fry are collected when only five days old.

The European system of breeding common carps is the Dubsich method. This method uses small ponds, often squarish, measuring about 8 × 8 m, where grasses are grown on the bottom to a height of about 40 cm when the ponds are dry. Dubisch ponds have a peripheral ditch, about 40-50 cm deep, flanking the edges of the pond, which serves as a haven for broodfish and fry. If grasses not grown in the pond, a special grass-vovered spawning board is placed in the pond center. In a variant of the Dubisch pond called the Hofer pond, instead of a peripheral ditch, there is a harvesting ditch on one side of the pond. Broodfish are removed from the Dubisch pond after they have spawned. The fry are transferred to prepared nurseries for further rearing about seven days after spawning.

In the Chinese system of breeding common carp, a modification of the Dubisch method is adopted using weeds like *Eichhornia, Ceratophyllum* and *Myrophyllum* as egg collectors. To prevent vegetation from drifting away from the spawning area, a squarish bamboo frame is use to contain the weeds.

In India, rarely is a natural pond directly used for common carp breeding. The most commonly used contrivance is a breeding *hapa* (top lid stitched at all sides except one broadside for release of vegetation and spawners inside) in which aquatic vegetation like *Hydrilla* and *Najas*, at 2 kg thereof per kg weight of female spawner are uniformly spread to serve as egg collectors. At times kakabans made, from coconut fibers are used.

Hatching in India done in hatching *hapas* into the inner wall of which weeds with eggs sticking to them are released. After hatching, the hatchlings pass through the holes of the inner *hapa* and collect themselves into the outer *hapa*.

Controlled Breeding of Common Carp

A measure of control on breeding of common carp is obtained by resorting to hypophysation, a system relevant to the species which was developed by Woynarovich (1969). According to this system, ripe female common carps are first anesthetized by inserting in the mouth sausage made with cotton wool dipped in MS222 (0.5 g/50 ml water), and the vent is sutured with waxed cotton thread to prevent inadvertent discharge of eggs when operators are not ready to receive them. Homoplatic pituitary extract at 2.5 to 3.7 mg/kg weight of female spawner, in a normal saline medium thickened with glycerine (70:30) ratio saline: glycerine, is injected following the same procedure as for other carps in respect of both sexes. In unsutured fish natural ovulation may be expected eight hours after injection at 28° C. Stripping is used to obtain fuller evacuation of the ripe ova.

For proper handling of common carp eggs, it is necessary to remove the adhesive substance which make the eggs of the species sticky. For this, a solution is made of 10 liters. Of distilled water, to which 30 g of urea and 40 g of sodium chloride are added. This solution is divided into two equal parts. One-half is gently added to the egg-milt mixture with continuous stirring for 3-5 minutes, the other half is added in small quantities at the time, at intervals of 5 minutes for 1-1/2 hours. The eggs swell up and become water-hardened in the process. To finally remove the adhesiveness of eggs, their entire mass is treated with a solution of tannin in water and the treatment repeated several times. The first treatment is with a solution of the strength 15g/10 liters of water. One-and-a-half to 2 liters of such a solution is added to the eggs for 10 seconds. The solution is then rapidly removed. This process is repeated with the tannin solution getting progressively weaker. After each 10 second treatment with tannin solution, the eggs need to be washed with pure tap water since tannin is toxic to eggs if left in contact with them for prolonged periods.

Sechoonbee and Prinsloo (1984) compared methods for removing carps egg adhesiveness and concluded and a recent improvement of the urea/NaCl method in which the urea concentration is raised to 20g/ liter is most effective (Woynarovich and Woynarovich 1980). An alternative method is to use a suspension of full cream powdered milk at 12-24 g/liter to rinse the eggs. Small fat globules form around the eggs and prevent them from clumping (Sechoonbee and Brandt 1982).

Hatching and Care of Hatchlings

Naturally spawned and fertilized eggs of carps, be they common carps or any of the species of Chinese or Indian major carps, if allowed to hatch in stagnant ponds are normally subject to heavy predation and exposure to disease due to infections and physico-chemical hazards of environment such that the percentage hatching may be less than even one. For this reason, fertilized eggs are generally collected after water-hardening and removed to special incubating devices for hatching. Hatching and the care of hatchlings and larva rearing, in fact, constitute the most important aspect of hatchery management and can make all the difference between success and failure of the operations.

A measure of protection from predation is afforded by installing a traditional hatching *hapa*. In its simplest form, a hatching *hapa* is a double-walled contrivance in which an inner *hapa*, made of a mosquito netting type of material, is installed within an outer one made of whole cloth. Obviously the later, which has to contain the inner *hapa* is the larger of the two. A *hapa* is generally fixed to the pond bottom by bamboo poles. Floating *hapas* have also been devised.

The advantage of having the inner *hapa* of mosquito netting material is that it helps separate the egg shells from hatchling. As soon as hatched, the hatchlings wiggle through the mesches of the netting leaving the egg shell on the outer upper surface of the inner *hapa*. The hatchlings accumulate in the outer *hapa* made of whole cloth. For best result in the *hapa* system of rearing, the breeding as well as the hatching *hapas* should be installed in running water, and predation from birds and frogs cut out by providing a cover on the top.

A pond hatching device, like a hatching *hapa*, suffers devastating consequences from a rise of water temperature as well as a drop of water level. Either can lead to total mortality. A running water supply in a flow-through system, where the rate of flow of water, its temperature and dissolved oxygen content can be controlled, and treatment to control bacterial an fungal infection given are requisites of an efficient carp hatchery.

The various types of hatching apparatus and device which can be installed in an indoor hatchery, taking care that clean oxygenated water of the right temperature at a flow rate of 1-2 liters / minute is maintained in the flow-through system. If different systems of hatching, such as jars and circular tanks, and available for use in the same hatchery, then the choice as to which one to use should depend on the magnitude of the breeding operation to be done. Normally, jars may be used for small-scale operation and circular cement tanks for larger-scale ones. The volume of fully swollen and water-hardened laid eggs of different species may be used in deciding the number of jars of hatching tanks that are to be put to use assuming enough eggs are added to fill the jars up to about ½ their individual capacities.

After the eggs have hatched, the hatchlings may be transferred to larvae rearing tanks or kept in a cloth *hapa* for about four days, during which period they need no exogenous food since they subsist on their own yolksacs. Figure 6.13 shows carp hatchlings with their yolksacs. The hatching jars or the circular breeding tanks must, however, be vacated to provide room for the next lot of eggs and / or broodfish. It normally takes slightly more than three days for the yolksac to be absorbed. During the period when hatchlings are laden with a gradually diminishing load of yolk, they alternate between periods of activity and inactivity, and their movement are vertical. During this period, the mouth of the hatchling is developing but not yet formed.

The hatchling must be given appropriate food from an exogenous source a little before the total absorption of their yolksac.

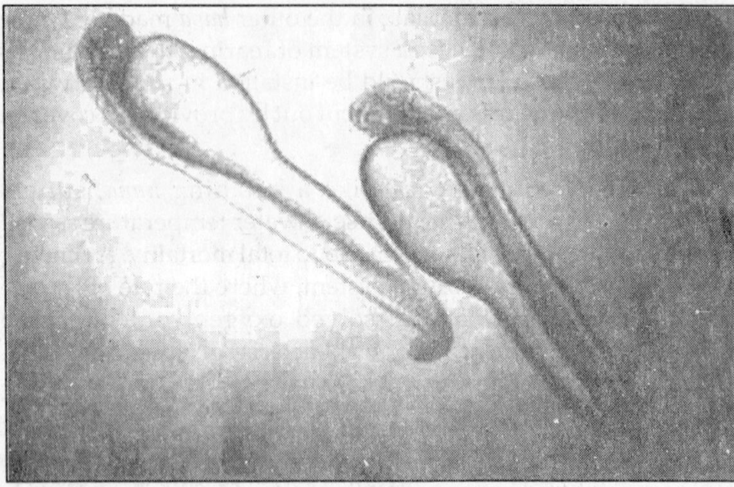

Figure 6.13: Carp Hatchlings Shortly After Hatching Showing the Yolksac

In order to test whether or not the hatchlings are ready to take exogenous food, on the third day after hatching, samples of hatchlings should be taken out and their behaviour watched in a glass beaker or deep petri dish. The vertical movement of the hatchlings changes into horizontal swimming movement when most of the yolksac is absorbed. As soon as that happens, the hatchling must be given live and/or artificial food. At the Freshwater Fisheries Development, Training and Extension Station (Balai Budidaya Air Tawar), Sukabumi, West Java, Indonesia, a mass culture system has been developed for rotifers. Rotifers are grown in vigorously aerated water fertilized with dried chicken manure (400 g/m^3) and treated with organophosphate pesticides (*e.g.* Fumadol, 0.25 to 3.0 ppm) to kill cladocerans and copepods. The rotifers are transferred to the post larval tanks by means of air-lift pumps. Figures 6.14 and 6.15 show the installation of air-lift system in the rotifer production pond. The organophosphate formulations Sumithion 50 EC3, Baytex and Dipterex in concentration 0.25 to 3.0 ppm have similar selective action on rotifers and crustaceans. Much more experimentation need to be done to really develop an effective system of sustained pure dense rotifer culture and to develop a method of their transfer to postlarval tanks. The Indonesian work has given brilliant indicative results.

Figure 6.14

Figure 6.15

Figures 6.14 and 6.15: Installation of the Airlift System in a Rotifer Pond at the Balai Budidaya Air Tawar (Development, Extension and Training Wing) of the Directorate General of Fisheries, Sukabumi West Java, Indonesia

If live food is not available, the first artificial food may be either microencapsulated whole chicken egg or boiled and mashed chicken egg yolk. The former being whole egg and being in microform, appropriate to the size of the newly formed mouth of the postlarva, is superior. The hatchlings may be kept in the hatchery on egg diet for one to two days, with feeding every two hours, after which they must be transferred to prepared earthen nurseries or otherwise taken care of for further rearing if not sold to aquaculturists customers. The nutritional values of the micro-encapsulated whole chicken egg and egg yolk diets are mentioned in Chapter 8.

Chapter 7
Postlarvae and Fry Rearing of Carps

Preparation and maintenance of nursery, rearing and stocking ponds are important steps in carp hatchery operations. Indian carp and exotic postlarvae and fry are delicate and unprepared or unmaintained ponds may lead to virtual decimation of the stocked material for various reasons stated further in this Chapter. The purpose of nursery, rearing and stocking pond preparation before stocking and maintenance after stocking is not only to remove the causes of poor survival, growth and health of the stocked material but also to optimize good husbandry factors for rearing the young of the species concerned.

Larval rearing in a carp hatchery has two distinct phases:

1. Rearing of postlarvae to the fry stage, usually in nursery ponds and
2. Rearing of fry to the fingering stage, usually in rearing ponds.

In exceptional cases, these steps are combined in one pond. However, instead of rearing postlarvae to fingerlings in the same pond continuously in two different types of water bodies, nursery and rearing ponds, as stated above. This is because there are differences between the postlarvae and fry in their food, stocking rates and environmental requirements.

However, factors which cause mortality, poor growth and ill health are, in great measure, common to both types of ponds. These are:

1. Physico-chemical incompatibility of the water of the hatchery proper and that of the nursery pond into which postlarvae are transplanted and likewise, disharmony of the waters of the nursery and rearing ponds (both need acclimatization in the respective ponds).
2. Lack of adequate amount of the requisite kind of fish food in nursery and rearing ponds.
3. Predatory aquatic insects other unwanted biota which naturally abound in all piece of freshwater bodies unless purposefully eradicated.
4. Predatory fish especially the young of catfish (*e.g. Wallago attu, Ompok bimaculatus, Clarias* spp. etc.) murrel (*e.g. Channa* spp.) and featherbacks (*e.g. Notopterus chitala*) may be present in the ponds and water supply sources.
5. Cannibalism among carp young, especially when postlarvae and fry of different size are co-stocked.
6. Sudden rise (or fall) of water temperature especially in cement or concrete tanks or earthen ponds with severe seepage problems accentuated when the water depth of the ponds is small.
7. Excessive growth of aquatic macro-phytes and phytoplankton and possible depletion of Dissolved oxygen particularly during nights and/or after several continuous rainy and cloudy days, causing asphyxiation of fish on the one hand and supersaturation of oxygen during hot sunny days causing gas embolism among fish on the other.
8. Inherent toxicity of certain algal blooms.
9. Afflictions by ecto- and endoparasites; bacterial, fungal and viral infections resulting in disease and mortality.
10. Abnormalities and ill health arising out of nutritional deficiencies.

To prevent the above maladies and create wholesome fish husbandry conditions, firstly, the nursery and rearing ponds should be designed to conform with the requirements such as drainage facilities depth provision of filteration system, since without these facilities the objective of pond preparation and maintenance cannot be achieved.

The next step comprises adoption of the measures in the sequence stated and at timings specified in the following account of nursery pond management. These measure fall into three categories:

1. Steps to be taken before stocking postlarvae in the nursery ponds;
2. Procedures to be adopted during stocking; and
3. Steps to be taken after stocking until the production of fry and their harvesting either for sale or for further rearing into fingerlings in rearing ponds.

Pre-Stocking Practices

Growing a Short-term Crop of a Leguminous Plant on the Pond Bottom

For preparing nursery and rearing pond before stocking, it is desirable to sow a short-term crop of a leguminous plant (peas, beans, clover, etc.) and after it is grown, to plough and level the pond bed so as to incorporate the plants with their roots into the pond soil. This process, called "green manuring", enriches the pond soil with nitrogen (most leguminous plants fix atmospheric nitrogen). This practice is believed to be beneficial for enhancing pond productivity by nitrogen fixation and enrichment and resulting in high survival and fast growth of stocked fish. However, the usefulness of nitrogen as an added fertilizer (which is different from growing leguminous plants) has been questioned because phosphorous and potash fertilizers given equally good results in enhancing productivity. These concepts are discussed further in the section of inorganic fertilization. There is no information available on the optimum density for sowing the leguminous plants or on quantities of legumes required to make an impact on pond productivity.

Application of Toxicants to Eradicate Unwanted Fish

This step in pond preparation should be taken about three weeks before the anticipated date of release of postlarvae.

A number of toxicants of plant origin can be successfully used for control of trash fish. Some of the toxicants of plants origin, besides initially acting as piscicides, at a later stage (after their initial toxicity has been used up to kill fish) also fertilize the water. Oil cake of the plant, *Bassia latifolia*, commonly called mahua oil cake in India, at a dose of 200-250 mg/L, gives a complete kill of predatory and weed fishes within 10 hours. Its application should be at least two weeks before stocking the ponds with carp hatchlings. The toxicity of mahua oilcake, at doses of 200-250 mg/L. may last for 72 to 96 hours. Tea-seed cake can also be used to rid ponds of wild fish, tadpoles and insects. For use a toxicant, the tea-seed cake is first ground, soaked overnight to soften and then applied at the rate of 525 to 675 kg/ha. Since tea-seed cake at this dose raise pond acidity, quicklime (calcium oxide) at a additional dose of 150 kg/ha should be used to remedy the situation. Where the purpose of applying lime is to serve as a general disinfectant or precipitation of excess dissolved organic matter, then it should be broadcast over the pond water surface.

For thorough disinfection of a pond, dose of 1,000 kg/ha of quicklime is required, but liming has been regularly done annually in the past, 100-200 kg/ha are enough unless the pond soil is very acidic or very poor in carbonates.

Quicklime along can be used in place of a toxicant to kill wild fish, insects and tadpoles and additionally bestowing on the pond the other benefits of liming. For this purpose, a dose of 900 to 1,050 kg/ha of quicklime is necessary if there is little water in the pond. If the pond is full, the dose of quicklime should be increased to 1,575 to 2,250 kg/ha.

Successive treatments with quicklime and tea-seed cake, the former in the dose of 1,575 to 2,250 kg/ha and the latter 525 to 675 kg/ha. Followed by a second quicklime treatment at the rate of 150 kg/ha given most effective results in killing unwanted pond biota, including predatory and weed fish as well as in giving the benefits of liming.

The calcium requirements of pond soils are best determined by soil pH, as seen in Table 7.1.

Other substance when are of tested and proven value is fish toxicants are powdered weed kernel of *Croton tiglium* (3-5 mg/L) powdered roof of *Milletia pachyacarpa* (2-6 ml/L), powdered seed of

Berringtonia accutangula (20 mg/L) powdered unripe fruit of *Randia dumetorum* (12 mg/L) and powered bark of *Walsura piscidia* (10 mg/L).

Table 7.1: Pond Soil pH and Requirement is Hundreds of kg/ha of Different Forms of Calcium

Soil pH	*Calcium Oxide/ha*			*Calcium Carbonate/ha bottom Pond Muds*
	Clay or Loam		*Loamy Sand*	
	Loam	*Sand*		
4.0	40	20	12.5	60-120
4.0 to 4.5	30	15	12.5	48-96
4.5 to 5.0	25	12.5	10.0	36-72
5.0 to 5.5	15	10	5	30-48
5.5 to 6.0	10	5	2.5	16-30
6.0 to 6.5	5	5	0	14-16

After Hora and Pillay (1962).

Liming

The advantages of liming a pond are numerous. In general liming enhances pond productivity and improves its sanitation. It is both prophylactic and therapeutic. Specific advantage are that liming.

1. Kills pond bacteria, fish parasites and their intermediate life-history stages, hence, especially efficacious in a pond where there has been outbreak of an infectious disease.
2. Builds up alkaline reserve and effectively stops fluctuations of pH by its buffering action.
3. Renders acidic waters usable for aquaculture by raising their pH to alkaline levels.
4. Neutralizes iron compounds which are undesirable to pond biota including fish.
5. Improves pond soil quality by promoting mineralization.
6. Precipitates excess of dissolved organic matter and thus reduce chances of oxygen depletion.

7. Acts as a general pond disinfectant for maintenance of pond hygiene.

The commonly available and use forms of lime are calcium carbonate (ground limestone), calcium hydroxide (slaked lime) and calcium oxide (quicklime). Calcium carbonate dissolves slowly and is of special value for pond fertilization and building up alkaline reserve also leading to calcium enrichment. Quick rapidly binds acids and influences pH rapidly, producing results similar to other forms of lime in half the quantity.

Lime can be applied to the pond bottom, added to water at inlets or uniformly broadcast on the water surface depending on the form of lime and the purpose of application. Calcium hydroxide and calcium oxide are best applied on the pond bottom after it has been drained.

Pond Fertilization

The next step in nursery preparation is fertilization, the objective of which is the sustained production of adequate quantities of zooplankton which form the natural food of carp hatchlings and fry.

In the early days of carp culture, the production in a nursery of any species of zooplankton, whether a cladoceran, a copepod or a rotifer or a protozoan regardless of size, was considered adequate. However, in recent times it has become clear that the postlarvae of carps survive and grow best if they are fed on smaller planktonic forms like free-living protozoa and rotifers; and that fry and fingerlings, whose mouths are bigger than those of postlarvae, grow best if feed on larger planktonic organisms like cladocerans and copepods. In fact, it has long been known that some species of copepods prey on carp larvae and postlarvae. Methods of producing pure or nearly pure cultures of rotifers are dealt with later in this Chapter.

For production of zooplankton nursery ponds are treated either with organic manures (such as cattle, pig or chicken manure) alone and/or with inorganic fertilizer such a NPK mixture. If both organic manure and inorganic fertilizers are used they may be applied either one following the other or as a mixture.

Organic Manuring

If animal manure is to be used alone, its dose should depend on the fish toxicant used for the eradication of unwanted fishes. If the fish toxicant used is mahua oilcake or tea-seed cake, which as stated earlier, have manurial value, the pond is initially manured generally with only 5,000 kg/ha of dry or near dry cattle manure about two weeks before the anticipated date of stocking following by repetition at the same dose seven days after stocking. But with other toxicants, which have no manurial value, cattle manure should be applied at a initial dose of 10,000 to 15,000 kg/ha about two weeks before the anticipated date of stocking followed by further manuring at the rate of 5,000 kg/ha seven days after stocking. These manurial rates produce enough zooplankton for a single crop of postlarvae stocked at 1.5 million/ha. If two or more crops of fry and to be produced from the same nursery pond, then the pond should be fertilized with 2,000 kg/ha of cattle dung about a week before each subsequent stocking. If poultry manure is to be used instead of cattle manure, one-third the dose of the former is sufficient since it is at least three times richer in their nitrogen, phosphorous, potassium and calcium than cattle or pig manure.

If organic manures are not available, then commercial compost, as used for agriculture, may be applied as a substitute. The recommended dose for compost is 5,000 kg/ha two weeks before the anticipated date of stocking followed by 5,000 kg/ha a week after stocking. If however, compost is to be produced in the carp hatchery itself, then the procedure to adopt is to dig the requisite number of pits about 4 m × 3m × 2.5 m deep at an isolated location and dump green vegetation in heaps about 30 cm high alternating with 7.5 cm high layers of cattle manure, both dusted liberally with calcium superphosphate. A ratio of 10:1 carbon to nitrogen is required for rapid decomposition of vegetable matter in composting. For this purpose, 25 kg sodium nitrate (per 100 kg of compost) should be applied with 4,000 liters of water to provide the necessary humidity in the compost pit. The compost should be turned initially three weeks after dumping and thereafter after every five weeks again. Compost may be expected to be ready in a total period of about 12 weeks after filling pit at temperatures ranging from 18°C to 35°C.

Water hyacinth (*Echornia crassipes*) has become a widespread menace in almost all the countries of South and Southeast Asia in recent decades, virtually choking wide areas of still-water masses, canals, drains, etc. The weed can be utilized for composting for use in fish ponds. For composting, 60 per cent water hyacinth, 38 per cent cattle manure, 1 per cent lime and 1 per cent urea, can be dumped in layers one above the other in the compost pits. As an alternative, stacks may also be set up above the ground. In that case, the stacks should be kept covered with plastic sheeting or soil all the time. Water hyacinth compost can be prepared in about 30 days.

Inorganic Fertilization

There is so far no generally accepted method of using inorganic fertilizers in ponds. There is usually a need for nitrogen and phosphorus and (less frequently) potassium which can be made good individually be applying sodium nitrate, super-phosphate and potassium chloride, respectively. Generally, however commercial fertilizers containing known amounts of available nitrogen, phosphorus and potassium are used to make good deficiencies of more than one element. Many combinations of these element are available as NPK[6] soil fertilizers *e.g.*, NPK 6:8:4 is widely available as a NPK, 16:20:0. NPK 18:8:4 is common in India. Then main point is that these mixtures were developed for soil, not water, fertilization. No one has yet worked out critically the mixtures required for pond use. The N:P ratio is the most important aspect-K is far less likely to be limiting. Moreover N and P source is the pond soil are generally less important than the exogenous amounts needed for good water column fertilization although soil source are very variable (*e.g.*, Table 7.2). Recent research suggest that a 10:1 elemental N:P ratio is required for phytoplankton. However most soil fertilizers give a much lower ratio. Therefore Swingle (1947) recommended adding 11 kg/ha sodium nitrate to raise the N in his 6:8:4 NPK mixture (applied at 45 kg/ha to 7:6:8:4) in Alabama ponds. Beyond such general guidelines there are no clear guidelines. Inorganic fertilizers usually applied in 10 equal monthly installment at 100-150 kg/ha/yr.

Pond Fertilization with Mixed Organic and Inorganic Fertilizers

A mixture of organic and inorganic manures applied at the following rates has given good results, cattle manure at the rate of 20,000 – 25,000 kg/ha per year and inorganic fertilizer mix (ammonium sulphate + single superphosphate + calcium nitrate in the ratio 11:5:1) at the rate of 1,380 – 1,725 kg/ha per year applied in 4-10 equal installments spread over the year.

Table 7.2: Classification of Indian Soils Based on Available Phosphorus and Nitrogen (mg/100 mg of soil[a])

	Available Phosphorous (P_2O_5)	*Available Nitrogen (N)*
High	6–12	55–75
Medium	3–less than 6	25–less than 50
Low	0–less than 3	0–less than 25

a: Khan and Jhingran (1975).

Fertilization of ponds with organic and inorganic substance may sometimes result in harmful dense phytoplankton blooms. This is traditionally controlled in India by sprinkling liquid cowdung (at the rate of 100 kg/ha) or some other organic matter or by covering the surface of the pond with the duck weed *Lemna*, causing the bloom to die within 4-5 days by preventing penetration of sunlight. Some innovative fish farmers use dense floating mat of water hyacinth (enclosed in a bamboo framework making it a floating island) to attain the same objective. However, these are empirical practices. Wet cowdung application can lead to excessive rise of BOD with tragic consequences. Chemical methods of phytoplankton bloom control should be used and are mentioned further in this Chapter.

Use of Land Vegetation as Pond Fertilizer

If organic manure and inorganic fertilization are not to be used land weeds, green grasses and vegetable wastes (especially those with soft stems) make good material for vegetation-based pond fertilizer in nursery ponds. The grasses, at 22,000 kg/ha, are enclosed in a rectangular bamboo-work frame in the nursery pond about a meter away from the bank and allowed to decay (Figure 7.1). The grasses are turned over in the frame 7-10 days later. Such grasses as have not decomposed and have remained hard and removed.

Subsequent fertilization with decayed vegetation has to be done by additional green vegetation at the rate of 100 to 130 kg/ha every week even after stocking. This is done as often as the pond condition permits, especially the dissolved oxygen level, which may be self-indicated by fry surfacing for gulping in air. Dissolved oxygen

Figure 7.1: A Rectangular Bamboo Frame for Enclosing Decaying Land Vegetation to Serve as a Pond Fertilizer (Reproduced from a Filmstrip Entitled "Rearing fry and fingerlings of Chinese carps," FAO)

should be chemically determined from time to time when organic manuring is being used.

Eradication of Predatory Insects and Other Harmful Biota

Nursery ponds more generally than not abound with a large number of aquatic insect over the greater part of the year, especially during and after rains. The more harmful one preying on carp spawn belong to the insect families Dytiscidae, Gyrinidae, Hydrophyilidae, Notonectidae, Plenidae, Nepidae, Belostomatidae, Corixidae, Gomphidae, Aeshnidae, Agrionidae and Coenagrionidae.

Treatment to be given to nursery ponds for insect control may be any one of the following :

1. Spraying an emulsion of 56 kg mustard or coconut oil and 18 kg washing soap per hectare, 12-24 hours before stocking the postlarvae.
2. Spraying an emulsion (per hectare) of 56 kg of mustard oil and 560 ml of Teepol, a detergent synthesized by Burmah Shell.
3. Spraying a mixture of oil extracted from the plant *Calophyllum inophyllum* and *Hertex*. W.P., a water

dispersible grammexane. This mixture, which is economical to use, is also very effective against the prawn *Macrobrachium lamerri*, which often abounds along with insects in nurseries. Further the mixture does not affect fish postlarvae and zooplankton. The insect kill may be expected within ½ to 11 hours depending on the concentration of *Hertex*. W.P. which may vary from 0.6 to 1 ppm.

4. Spraying, in the dose of 0.01 ppm, pure gamma isomer of benzene hexachloride dissolved in ethyl alcohol which is highly toxic to insects.
5. Spraying an emulsion of high-speed diesel oil (1 liter), the emulsifier Hyoxid 0.011 (0.75ml) and water (40ml) at the rate of about 1 liter of the emulsion for every 200 m^2 of water surface.
6. Spraying an emulsion of high-speed diesel-boiler oil with any available detergent.
7. Application of 0.25 to 3.0 ppm of the organophosphate substances Fumadol or Sumithion or Baytex or Dipterex. These substances are highly toxic to insects and are believed to be biodegradable. The substance Sumithion at 3 ppm, while being toxic to insects, is selectively toxic. Rotifers are virtually immune to it while insects, copepods and cladocerans are not.

Fairy shrimps often appear in large numbers in nursery ponds. If liming has not killed them already, additional liming at 4,000 ppm will eradicate them.

Certain birds, notably cormorants, kingfishers and herons when present are destructive to postlarvae and older fish. Thin lines stretched across the pond are the most effective means of controlling them.

Stocking Carp Postlarvae in Nursery Ponds

Judging Suitability of Environment Before Stocking

After controlling all factors which might lead to mortality of the stocked carp postlarvae, as described in the preceding section, the most appropriate time for stocking a nursery pond is when it abounds with zooplankton, especially rotifers and cladocerans in adequate density (preferably rotifers only). Before actually stocking the nursery pond with postlarvae, it is necessary to make an estimate of the types and abundance of plankton present. This can be done by adopting appropriate, limnological methods. A rough field method for estimating plankton has been developed by Alikunhi (1956) for use by fish farmers. In this method, about 55 liters of water, taken from different sections of the nursery pond, are filtered through an organdy or muslin ring net with a 2.5 cm diameter glass specimen tube tied to the lower end of the net. A pinch of powdered common salt is added to the water in the tube after the plankton is collected and the tube detached from the net. Within 15-20 minutes of adding the salt, most of the organisms settle on the bottom. If the column of plankton sediment is at least 15 mm high from the bottom of the tube and the sediment found to consist mostly of cladocerans and rofifers, the pond may be considered sufficiently rich in plankton to stock at the rate of 1.5 million postlarvae per hectare. The animal or plant nature of plankton sediment is roughly indicated by either a pale-brownish or greenish colour of the sediment, the former indicating preponderance of zooplankton and the latter of phytoplankton. If the predominant plankton population be not of zooplankton, then further organic manuring should be done at doses recommended to rectify the situation before stocking.

Method of Stocking

To avoid any abrupt change in quality and temperature between the water of the hatching tank and that of the nursery pond, the postlarvae should be kept in a suitable container having water initially from the former (hatchery tank) to which the water from the nursery pond should be gradually added in stages, eventually substituting almost the entire hatchery water of the container by the water from the nursery pond. The container should then be slowly dipped and tilted in the nursery pond so that the postlarvae are free and swim out of the container into the nursery pond. Stocking should be done late in the evening, a procedure which gives postlarvae time to acclimatize themselves during the ensuing night relatively free from depredations of enemies, should any have escaped the steps adopted to eradicate them.

Stocking Rate of Nursery Ponds

The stocking rate of postlarvae in a nursery pond depends on the management practices intended. If natural food the form of zooplankton is to be produced by continued pond manuring and supplement feeds are also to be given, and if facilities exist to remedy oxygen deficiency condition should this occur under condition of heavy stocking, then the stocking rate may be as high as 10 milion per hectare.

Post-Stocking Practices

Fish Feeding

Feeding, described in this Chapter, is based on natural live food organisms generated in the pond itself and augmented through fertilization and on supplemental feed given exogenously. This is a great deal of difference between complete fish feeds and only supplemental feeds.

Soon after being stocked in manured nursery ponds containing rich zooplankton, carp postlarvae start grazing voraciously on natural food. At this time the feed requirements of spawn are so large that within two to three days of stocking, the plankton initially present in the ponds gets exhausted and steps must be taken not only to generate more natural food but also to administer artificial feeds. The natural zooplanktonic food in the pond is increased by continued manurial treatment already described in a previous section in this Chapter. Supplemental feeding and manurial pond enrichment, when done simultaneously lead to high survival and fast growth of the stocked postlarvae in nurseries. The commonly administered artificial feeds for common, Chinese and Indian major carps are feed on rice bran and oilcakes of groundnut, coconut, mustard, etc. Artificial feeds are always given in finely powdered form to carp postlarvae.

The rate of administering supplemental feeds to carp hatchlings for 15 days of rearing is shown in Table 7.3. The proportion of different kinds of feeds given to carp postlarvae is shown in Table 7.4.

It has been found that among the combinations of various artificial feeds containing protein, fat, carbohydrate, minerals, roughage and vitamin, etc., the maximum growth of postlarvae is obtainable with feeds having a combination of hyrolyzed proteins and carbohydrates (50:30). Complex proteins and pure carbohydrates give poorer results. Further, rice bran alone, a food most often given singly, gives much poorer results than a mixture of oilcakes, rice powder and black gram in powdered form. Silkworm pupae and fishmeal give still better results.

Table 7.3: Feeding Schedule of Carp Postlarvae Fry for 15 Days of Rearing

Days after Stocking	*Feed/Day Expressed as Weight in g or Hatchlings at the Time of Stocking*
First 5 days	Equal to double the weight
6th to 10th day	Double to three times the weight
11th to 15th day	Three times to four times the weight.

After Alikunhi (1957); Hora and Pillay (1962).

Table 7.4: Kinds of Feeds given to Carp Postlarvae

Days After Stocking	*Kind of Daily Feeds as "Cups"*			*Rice Bran*
	Mustard Oilcake	*Groundnut Oilcake*	*Coconut Oilcake*	
2nd to 5th day	1.0	1.25	2.0	2.0
6th to 10th day	1.5	2.50	3.0	4.0
11th to 15th day	2.0	3.75	4.5	6.0

*A measure called Bati in India for this purpose equivalent to 130–150 ml.

A feed compounded from dried and finely powdered and sieved notonectids (which are highly predatory to carp spawn), small prawns and shrimps and cheap pulses or lentils in the ratio of 5:3:2 gives much better results for survival and growth of hatchlings of catla, rohu, mrigal and silver carp than the conventional mixture of rice bran and oilcake, besides utilizing an otherwise wasted product.

For monoculture of rohu spawn, good results are obtainable with zooplankton, followed by silkworm pupae, mustard oilcake with rice bran and groundnut cake with wheat bran in the order stated.

Normally postlarvae stocked in nursery ponds attain a length 2.0 to 2.5 cm (when they are termed fry) in about 15 days with artificial feeding giving a survival of about 50 per cent in earthen ponds.

A higher survival rate may be expected if a system of intensive controlled culture of rotifers and crustaceans to serve as fry food is developed and facilities for intensive aeration are provided. The fry may be transferred to prepared fry-rearing tanks or otherwise taken care of for further rearing if not sold to fish farmers.

Rearing and Pond Management

The fry are netted from the nursery ponds and released in rearing ponds which are to be prepared in the same manner as the nursery ponds. The tasks of preparing rearing ponds included : (1) elimination of predatory and weed fishes; (2) manuring with organic and inorganic fertilizers; (3) weed control; (4) stocking with carp fry; (5) supplementary feeding and (6) harvesting. The processes of elimination of predatory and weed fishes and manuring with organic and inorganic fertilizers are identical to those described in the case of nursery ponds. A description for achieving weed control follows.

Weed Control

Being somewhat deeper and larger than the nursery ponds, rearing ponds are more liable to get infested with weeds. An overgrowth of weeds depriver the ponds soil of nutritive elements, restricts the movement of fish, interferes with netting operations and harbors predatory and weed fishes and insects. Weed occupying different habitats and niches have to be controlled in different ways.

Floating weeds like *Eichhornia* and *Pistia* are best removed by manual labour. Chemicals like 2, 4-D are quite effective are economical against *Eichhornia*, though not frequently against *Pistia*. When mixed with common domestic detergent, 2, 4-D acts effectively against weeds like *Pistia*, *Nymphaea* and *Nelumbo* in which leaves are either hairy or waxy. Simazine WP-50, applied at 5.6-11.2 kg/ha kills *Pistia*, *Eichhornia* and *Colocasia* completely within two to three weeks even during rains. Taficide–80, at a dose of 2.2 kg/ha, is also effective against *Eichhornia*.

Marginal weeds like *Typha*, grasses, sedges, rushes, *Ipomoea*, *Jussiaea*, *Sagittaria* and *Colocasia* are effectively controlled by ploughing in, grazing by livestock, burying during dry season or repeated cutting and deepening or marginal shelves.

Rooted emergent weeds like *Limnanthemum*, *Trapa*, *Myriophyllum* etc. are successfully removed by repeated cutting of leaves before fruiting at weekly intervals, for about six to eight weeks. Alternatively, spraying once or twice with, 2, SD-4 (at 5, 6-11.2 kg/ha) kills these plants.

Rooted submerged weeds are cleared by a number of simple, manually operated devices like bottom rakes, log weeders, metal spikes, with or without barbed wire attachment, forks, drag chains or bamboo poles fixed with cross pieces at their lower end followed by repeated netting with strong wire or rope nets. Other methods employed included shading by floating plant like *Pistia* or *Salvinia* for a period of 8-10 weeks or by creating algal blooms or algal mats by repeated fertilization with N:P:K fertilizers.

Some of the better known fishes that are used for biological control of weeds are the grass carp, *Ctenopharyngodon idella* and *Puntius javanicus*. Grass carp feeds most voraciously on *Hydrilla, Najas* and *Ceratophyllum*. It can also control infestations of *Ottelia, Vallisneria, Nechmandra, Utricularia, Trapa, Myriophyllum* and *Limnophila*.

Anhydrous ammonia gas, obtainable in gas cylinders, controls *Hydrilla, Najas, Wolffia, Nymphaea, Ottelia, Limnanthemum* and *Nelumbo* when injected in the subsurface layers with an applicator at 112-334 kg/ha or 6.9-19 ppm. Much of the ammonia applied enters the pond's production cycle by nitrogen enrichment resulting in phenomenal growth of plankton soon after weed clearance has been achieved with ammonia application (Ramchandran 1963). Recently Subramanian (1983) showed that releasing ammonia through calcium hydroxide and ammonium sulphate in the ratio of 1:1:8 does not need the use of an applicator.

Some blue-green algae, particularly, *Microcystis*, can from long persistent blooms which deplete oxygen and often cause fish mortality. Simazine at a dose of 0.5 to 1.0 ppm clears algal blooms and mats and brings about prolonged control without affecting production of other plankton and fish.

Stocking of Carp and Supplementary Feeding

For further rearing of carp fry to fingerlings, either monoculture of polyculture in any species combination may be carried out. The stocking density of fry measuring 25.4 mm to 37.8 mm weighing 0.15 to 0.75 g each may be 125,000/ha to 250,000 /ha. A practice recommended for healthy fry rearing is that the size of the fry in a rearing ponds at the time of stocking should be as uniform as possible. This is accomplished by first segregating the catch in holding pens, made of gunny or canvas bottom and by first segregating the catch in holding pens, made of gunny or canvas bottom and split bamboo side (Figure 7.2) supported on stakes. It is necessary to periodically clean the holding pen of debris, fish excreta, etc. which is done by lifting the bottom and removing the rubbish manually (Figure 7.3). Size grading is done by sifting fry through sieves of different mesh gradation made of split bamboo. Two sieves use in Taiwan in culture of Chinese carps are shown in Figure 7.4. The early and advanced fry gradations accomplished by sifting are shown in Figure 7.5. Fry attain the fingerling length of 100 mm to 172 mm in about three months. A survival of up to 80 per cent may be expected provided that fully complement of management measures for environment enrichment are adopted and that the fry have natural food as well artificial feed given to them. Artificial feeding may be on oilcake, mustard or groundnut or coconut and rice bran in the ratio of 1:1. The density of feeding in the first month may be equal to the initial total weight of fry stocked daily and in the second and third months, twice the initial weight of fry stocked daily. If grass carp is among the fry in the fish stocked then *Wolffia*, *Spirodela* and *Lemna* may be given as feed at 1,750 kg/ha, twice in a month during the period of rearing. Table 7.5 shows some feeds and feeding rates for Chinese carp fry and fingerlings.

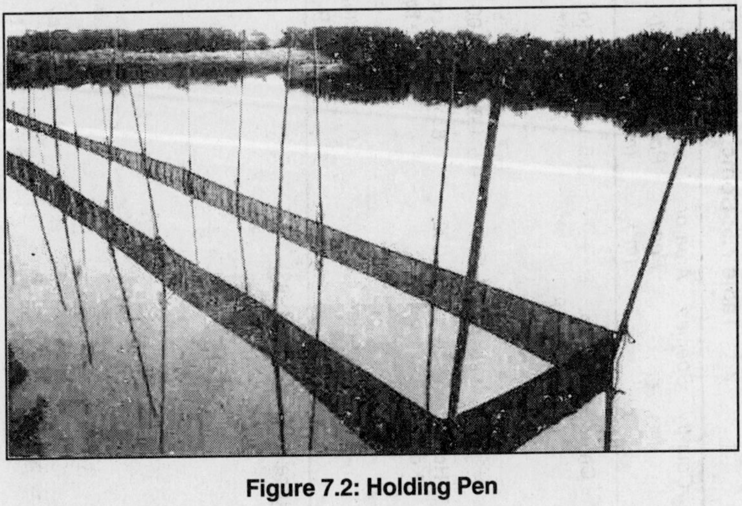

Figure 7.2: Holding Pen

Table 7.5: Some Feeds and Feeding Rates for Chinese Carp Fry and Fingerlings

Country	*Species*	*Area or Pond (m^2)*	*Depth of Pond (m)*	*Size of Fish*	*Age of Fish (days)*	*Stocking Rate (no./m^2)*	*Feed Rate*	*Feeding*
China	Bighead, grass carp	–	0.5-1.0	Up to 20 mm	Up to 30	100	Egg yolk paste or soybean milk, plus peanut cake after 10 days	1 egg/2,500-7,500 fry/day or milk from 300-500 g beans/50,000 fry/day
	All species	–	0.5-1.0	20-100	Up to 30	–	Soybean meal	45 kg/5,000 fry/month
Hong Kong	All species	1,000	0.8	0.8-3 cm 3 mg-1g	Up to 25-30	150	Soybean milk and peanut cake meal	100 kg soybean milk or 200 kg peanut cake meal/ month
	All species	1,400	1.0	3-12 cm; 1.15 g	30-70	35	Peanut cake, rice bran or soybean cake	Start at 1.5 kg/day, build up to 5 kg/day

Source: Chang *et al.* (1983).

Figure 7.3: Cleaning a Holding Pen

Figure 7.4: Sieves of Two Different Mesh Size made of Bamboo to Sift and Size-Grade Carp Fry before Stocking

Figure 7.5: Size-Grading Carp Fry by Sifting through Sieves

All figures are reproduced from a filmstrip entitled "Rearing fry and fingerlings of Chinese carp (FAO).

In China, bighead carp, grass carp and silver carp fry/ fingerlings are sometimes stunted during holding in nursery ponds. For a 1-ha pond 1.5m deep, holding one million, 4–100 mm fry/ fingerlings of any of these species the stunting diet is 0.5 to 2.0 per cent body weight of rice bran, soybean milk or peanut cake or alternatively 3-5 per cent body weight/ day of duck weed. In India also the Indian major carp young are sometimes kept stunted in West Bengal on naturally available food (without fertilization) up to almost a year and stunted fry and fingerlings are available on sale for commencing carp culture operations at any time of the year.

The empirical nature of pond fertilization and augmentation of natural food and young fish feeding with locally available material have already been mentioned. It is obvious that until complete young fish feeds are developed by commercial feed companies there can be no definite standardized feeding practices.

Harvesting

In the case of nursery ponds, harvesting may be done by repeated seining with fine meshed net. In rearing ponds, where fry are grown to fingerling stage, periodical harvesting may be done at an appropriate time to avoid over crowding. After harvesting, the fingerlings may be stocked unless sold to aquaculture farmers. The function of a carp hatchery is completed after fingerlings have been produced.

Chapter 8

The Nutrition of Cultured Carps

For carp culture in tropical developing countries, the feeding regimen is usually based on both natural and supplemental feeds. In other words, a nutritionally complete pellet, such as would be needed for high density, intensive culture in cages or tanks, is not usually required. In considering the application of nutritional data to carp culture, therefore, it is very important to recognize the differences between complete and supplemental feed formulation.

The common carp, in particular, has prospered as a cultured fish because of its omnivorous bottom foraging feeding habit and its good growth in fertilized ponds and on cheap feed which are truly supplemental to natural feeding. Gives the feeding habits of adult cultured carps. The formulation of complete feed for intensive culture is more difficult than that for supplemental feeds. Complete feeds usually contain generous amount of high quality protein, such as fishmeal, and are therefore, expensive (*see* Table 8.1).

A very important consideration in carp nutrition is the activity of intestinal bacteria which assist digestion and can provide nutrients supplemental to the ingested material. There role has not been fully investigated. This is a complicated factor in working out the dietary requirement of carps. However, sufficient data for the common carp now exist to have a broad picture of its protein, amino acid, lipid (including essential fatty acids), carbohydrate, mineral, vitamin and trace element requirements (Jauncey 1982).

Table 8.1: Composition of a 40 per cent Protein Carp Grower Diet (NRC 1977)

Ingredient	*International Feed No.*[a]	*Amount in Diet (per cent)*
Fish meal (mechanically extracted, 65 per cent protein)[b]	5-01-982	46
Wheat middlings, less than 9.5 per cent fiber	4-05-205	23
Rice bran with germ (solvent extracted meal)	4-03-930	7
Wheat bran	4-05-190	5
Soybean seeds (solvent extracted meal, 44 per cent protein)[b]	5-20-637	5
Torula yeast (dehydrated)	7-05-534	4
Corn glutein meal	5-02-900	1.5
Vitamin premix[c]	–	0.5
Mineral premix[d]	–	0.5
Sodium chloride	–	0.5
Potassium phosphate	–	2.0

a: Each feedstuff has a six-digit international number. The first digit gives the class of the feed. The systems for numbering are fully described in NRC (1977, 1983).

b: 6.25 × per cent nitrogen.

c: Vitamin premix : vitamins added to cellulose powder to make 0.5 per cent of diet (mg/kg).

Choline Chloride	500	Riboflavin	25
Ascorbic acid	80	Pyridoxine	8
Inositol	80	Thiamine hydrochloride	5
Niacin	60	Biotin	0.05
Calcium pantothenate	80	Vitamin A	8,000 (IU/kg)
Vitamin E	45	Vitamin D3	1,500 (IU/kg)

d: Mineral premix added to cellulose powder to make 0.5 per cent of the diet (mg/kg) :

Manganese	25	Magnesium	250
Iron	10	Cobalt	3
Zinc	25		

All the species mentioned in, including the specialist plankton feeders, will accept supplemental feeds. However, since data for the formulation of such feeds are confusing and also to some extent location-specific according to the natural feeding available, they will not be considering in detail here. Most supplemental feeds are made up by experience on an empirical basis. For example, the supplemental broodfish feed recommended for all carps in fertilized ponds at the Raipur hatchery, Bangladesh consists of 3-5 per cent body weight daily of a wheat bran, oilcake, coarse wheat flour, fish meal mix in the ratio 4:4:1:1. Even complete feeds are sometimes formulated on an empirical basis. For example, in cage culture trials with grass carp, a 50:50 commercial trout pellet : wheat flour mixture has been used successfully as a complete feed for all growout stages from < 10 g to 250-300 g. The composition of the feed was about 34 per cent protein, 48 per cent carbohydrate, 7 per cent fat, 2 per cent fiber and 10 per cent ash on a dry weight basis with a total energy content of about 4.4 kcal/g dry matter.

A complete guide to feed ingredient for warmwater fishes is provided by NRC (1977, 1983). For advice on the principles of supplemental feed and formulations, readers should also consult a guidebook prepared for tilapia culture which contains very useful summaries of feeding principles are practices (Jauncey and Ross 1982). This information, some of which is given below, can be applied in carp culture after reference to data on carp nutritional requirements.

Feed ingredients, particularly proteins and carbohydrates, must be digestible as well as sufficient in chemical composition. Table 8.2 shows that common carp are well able to digest plant proteins are most fish can digest animal and microbial proteins and a wide range of carbohydrates. However, the utilization by fish of plant cell wall material (celluloses and other structural materials collectively termed fiber) is largely a matter for conjecture. In the tilapias, it has been shown conclusively that blue-green algal cell wall digestion (and probably also, therefore, the utilization of detrital bacterial proteins) is highly dependent on a cycle of very strong acid secretion, stimulated by stomach filling. For stomachless carps, this mechanism cannot operate. There is an open field for further research here. In general, carp use dietary protein and lipids for energy in preference to carbohydrate (but see discussion below).

Table 8.2: Plant Protein Digestibility as Percentages by Ruminants, Swine, Rabbits and Common Carp (Reproduced from NRC 1977; Original source, Nehrin 1965)

Plant Protein Source	*International Feed No.*	*Ruminants*	*Swine*	*Rabbits*	*Common Carp*
Wheat, grain	4-05-211	91	–	–	83
Barley, grain	4-00-549	77	70	33	64
Rye, grain	4-04-047	83	–	–	63
Oat, grain	4-03-309	83	49	64	64
Corn, gent Yellow, grain	4-02-935	80	70	25	66
Peas, seeds	5-04-597	80	80	64	79
Soybean, seeds, heat processed	5-04-597	80	80	74	81
Lupine, sweet yellow, seeds	5-08-453	92	90	–	85

The Nutritional Requirements of Adult Common Carp and Other Cultured Carps

General Considerations

First, it must be emphasized again that the data available on the common carp greatly exceed those available for all other cultured carps. The common carp is best regarded as a benthic omnivore. It has no stomach and, like other carps, tends to feed for long periods of time. Therefore, in intensive carp culture, fish are usually fed small amounts of feed at frequent intervals. For example, the best feed conversion for 40g carp at 23°C is obtained from giving a total of 2.2 per cent body weight per day dry, complete pelleted diet as nine equal feedings. The best growth, however, is obtained using the same frequency at 6.5 per cent body weight per day. The known optimum growth temperature ranges for carps are: common carp, 23-30°C; sliver carp, 30-31°C; grass carp, 22-25°C. The Indian major carps prefer high temperatures, close to 30°C. Most experimental studies on the common and Chinese carps have been made towards the lower end of this range, and its should be remembered that feeding habits, feed conversion and nutritional requirement (particularly energy substrates) will be different at higher, tropical temperatures. Food passage through the fish gut quickens with temperature and can double in speed for a 10°C rise. For grass carp food passes through the gut in about eight hours at 30°C. Metabolic rate and growth are also faster at warmer temperatures and thus, energy substrates in the diet and protein for tissue synthesis are required in larger quantities. There is obviously a need for further study on the nutritional requirement of carps at tropical temperatures. At temperatures below the optimum ranges, growth fall off rapidly. For example, silver carp and grass carp growth poorly below 20°C, show very poor appetite at 10-15°C and stop feeding entirely below 9-10°C.

The second consideration is on protein requirements including essential amino acids. Most experimental studies have shown that carp based on high protein diets. For example, dietary protein levels of 28 to 35 per cent, with metabolizable energy content around 3.4 to 3.8 kcal/g, have been recommended for adults and broodstock. This may seem surprising for a benthic foraging fish, but it is not really so. A similar situation exists for the tilapias in which protein levels of around 40 per cent are recommended for intensive culture. This merely indicate that the feed consumed by microphagous/ detritivorous fish are exceedingly rich in proteins from various sources-bacteria, microalgae, meiofauna and small invertebrates.

Carp, like other fish species so far investigated, require ten essential amino acids: arginine, histidine, isoleucine, leucine, lysine, methionine, phenylalanine, threonine, tryptophan and valine (Table 8.3). Deficiencies in one or more of these can cause growth depression and possible deficiency syndromes which look like disease conditions, for example, skeletal deformities. The proteins take in by carps must, therefore, be sufficiently digestible and also supply sufficient amounts of all these compounds. Although carp intestinal bacteria can synthesize both essential and non-essential amino acids to some extent, this should not be relied on in formulating complete diets.

Basically, the data in Table 8.3 indicate that common carp has similar quantitative essential amino acid requirements to other cultured fishes, such as trout, salmon and eels. For example, the requirements of trout for the amino acid lysine and methionine (the two most likely to be limiting is processed feed components) are 2.12 (5.3/40) and 0.72 (1.8/40) using the same notation as in Table 8.3. This leads to the conclusion that the common carp requires not only high protein diets but also high quality animal or bacterial protein for optimum growth. Plant proteins alone are not sufficient. For example, soybean protein concentrate given at 30 per cent level as the sole protein source gives a methionine–deficient diet. A good summary on warmwater cultured fish nutritional requirements and extensive tables on the digestibility and composition of feed ingredients (including protein and amino acid profiles) are given in the reviews by NRC (1977, 1983).

Table 8.3: The Essential Amino Acid Requirements of Common Carp (*Cyprinus carpio*), Summarized from Jauncey (1982). Values are expressed as g/100 g of dry diet. In parentheses, the numerators are g/100g of protein and the denominators are percentage total protein. 1-data from Nose (1978); 11-data from Ogino (1980).

Amino Acid	*1*	*11*
Arginine	1.6 (4.3/38.5)	1.52 (3.8/40)
Histidine	0.8 (2.1/38.5)	0.56 (1.4/40)
Isoleucine	0.9 (2.5/38.5)	0.92 (2.3/40)
Leucine	1.3 (3.3/38.5)	1.64 (4.1/40)
Lysine	2.2 (5.7/38.5)	2.12 (5.3/40)
Methionine[a]	1.2 (3.1/38.5)	0.64 (1.6/50)
	Cystine = 0 per cent	Cystine = +
	0.8 (2.1/38.5)	
	Cystine–2 per cent	
Phenylalanine[a]	2.5 (6.5/38.5)	1.16 (2.9/40)
	Tyrosine = 0 per cent	Tyrosine =+
	1.3 (3.4/38.5)	
	Tyrosine = 1 per cent	
Threonine	1.5 (3.9/38.5)	1.32 (3.3/40)
Tryptophan	0.3 (0.8/38.5)	0.24 (0.64/40)
Valine	1.4 (3.6/38.5)	1.16 (2.9/40)
Total	13.7 (35.8)	11.28 (28.2)

a: Note that methionine requirements vary with the level of cystine present in the diet and phenylalanine requirements with tyrosine.

Fortunately, the protein requirements of carp can be modified by raising the levels of other dietary components. This called sparing the dietary protein component. The most important sparing compounds for common carp are lipids. These supply energy which would otherwise be used by metabolizing proteins and amino acids. For example, the optimum protein of young mirror carp feed in Hong Kong was reduced from 38 to 33 per cent; by adding 5 per cent soybean oil. The additionak increased dietary metabolizable energy from 2.8 to 3.1 kcal/g. Moreover, as common carp are generally accepted to be fatty fish, with carcass lipid levels as high as 20 per cent from pond culture, there is great scope for increasing dietary lipid content even it this results no growth retardation, no unacceptable carcass lipid buildup and improved protein utilization.

The data on sparing dietary protein in carp by other compound are confusing. Attempts to use nonprotein nitrogen compounds such as ammonium citrate and urea have not produced any clear results for application by culturists. The other obvious energy source for sparing dietary protein is carbohydrate. Carbohydrate requirements are discussed below. Attempts to raise dietary carbohydrate levels to spare protein have met with mixed results. Recent results, however, suggest that dietary energy as carbohydrate can be used by mirror carp; for example, fingerlings grow very well on a 45 per cent rice feed and protein digestibility is improved. The relationships between dietary protein and energy level and their effect upon carcass composition are complex. For further recent experimental result the reader should consult Zeitler *et al.* (1984).

Carps appear to be less sensitive to mineral-deficient diets than other fish. In pond fish culture, natural feeding on plankton and benthos can probably supply their dietary mineral requirements but, as in the case of vitamin, the inclusion of a mineral premix in supplemental and complete feed is advisable to make sure.

The two elements most likely to cause serious deficiency symptoms in carps are calcium and phosphorus, since these are required for bone and other tissue synthesis. Most commercial feeds contain sufficient calcium and phosphorus for carps in their fish-meal components. Moreover carp can utilized calcium from water, if present at 16-20 ppm. To make good dietary deficiency. However, phosphorus is present only in very low concentrations in most freshwaters and with supplemental feeds, in which fish-meal has been replaced by plant or other protein sources, deficiencies can arise. Carp fed on complete diets low in available phosphorus show increased lipid level in the viscera. Much of the total phosphorus in cereal is chemically bound to phytic acid. Fish cannot use phytin-bound phosphorus since they lack the digestive enzyme phytase. Japanese work has shown that the best growth of common carp is obtained on feeds containing 0.6 to 0.7 per cent available phosphorus. Deficient diets can be made good by adding 5 per cent monobasic calcium phosphate to the feed mix, but for cheap supplemental feeds based on locally available materials, culturists need only be sure that (1) the total feed mixture contains sufficient phosphorus and (2) that the phosphorus is available. Note that the availability of phosphorus to common carp is not the same as for fish which have a stomach and acidic digestive system (see Table 8.7).

Hepher and Sandbank (1984) have shown that phosphorus supplementation in common carp feeds can result in increased growth in some circumstance. For example, 1 per cent di-calcium phosphate supplementation is recommended for large fish in polyculture systems where the standing crop of common carp is high (>1.8 ton/ha) and phosphorus from natural feed may be inadequate.

The Feeding Habits and Nutritional Requirement of Carp Postlarvae, Fry and Fingerlings

Very little detailed work has been done on the feeding habits of carp postlarvae, fry and fingerlings. However, it is known that common Chinese and Indian major carp all require a microzooplankton diet initially, feed on progressively larger items as they grow in size and make a gradual transition to the appropriate adult feeding habit as fingerlings (see Table 8.8 for examples). The grass carp is the most striking example, changing gradually from microzooplankton to larger zooplankton to a more diversified benthic and planktonic diet and finally switching to herbivory as the pharyngeal teeth develop.

A further common feature of the early life history of all the culture carps is their rapidity of growth and development from small hatchlings (5-8 mm; 2-5 mg) to large active postlarvae and fry. Common carp hatchlings can resorb most of their yolksacs within 24 hours of hatching at 25°C. Moreover, fry can reach 10-15 mg within 5 days and 50-60 mg within 10 days. The growth requires large food intake (Table 8.9). Carp fry are exceptionally good at capturing living and non-living food items.

Hatchery/nursery feeding practices are largely dependent on the provision of live food organisms such as brine shrimp (*Artemia*) nauplii, rotifers, caldocerans (such as *Bosmina, Daphnia* and *Moina* spp.) and mixed plankton. Sometimes as monoculture of a food organism is prepared (for example, *Artemia nauplii* or *Moina micrura*) and sometimes a natural mixed population is allowed to develop (for example, mixed rotifer culture by fertilizing tank water with manure bags). In addition to the provision of live food organisms, supplemental feeds such as the microencapsulated egg diet, yeasts and proteinaceous flours have also seen some use. These are not considered in detail here, apart from the egg diet, since their use has not been standardized. Moreover, commercial feed companies are now beginning to market carp starter feeds.

All these feeding practices are covered below, but it should be recognized that they are largely based on empirical ground, not on any critical assessment of nutritional requirements. In fact, current feeding practices often constitute what may be termed a nutritional overkill. This is acceptable and even desirable for feeding fry and fingerlings since the amounts of feeds needed are relatively small compared to growout feeding requirements. Least cost formulation of feeds is, therefore, less important in hatchery/nursery work than in growout in all aquacultural operations. However, the early rearing of carps on defined artificial feeds under more controlled conditions remains a worthwhile goal. This is because the various combinations of live and artificial feeds and fertilization procedures in current use can cause unpredictable environmental consequences such as bloom of opportunistic plankton.

Species and Algae

The approach of use green water in hatchery work is widespread in aquaculture and as usually very successful since the phytoplankton effectively condition the water, lowering ammonia levels and increasing the dissolved oxygen. However, undesirable consequences such as blooms of nutritionally deficient plankton and increased populations of harmful ciliate protozoans in tanks and predatory crustaceans and insect in ponds can also occur. Trends towards better defined feed and feeding practices in early rearing are discussed below.

Table 8.4: Food Organism Preference of the Early Life History Stages of some Chinese Carps (Modified from Chang *et al.* 1983)

Stage/Size Range (mm)	*Grass Carp*	*Silver Carp*	*Bighead Carp*
Hatchlings (7-9)	Protozoa, rotifers, copepod nauplii	Protozoa, rotifers copepod nauplii	Protozoa, rotifers, copepod nauplii
Fry (10-12)	Protozoa, rotifers, small cladocerans and copepods	Protozoa, rotifers small cladocerans	Protozoa, rotifers, copepod nauplii, small cladocerans
Fry (13-17)	Cladocerans, copepods, microzoobenthos	Protozoa, rotifers, small cladocerans	Protozoa, rotifers, copepod nauplii, small cladocerans
Fry (18-23)	Cladocerans copepods, microzoobenthos, Chironomus, detritus	Protozoa, rotifers small cladocerans, phytoplankton	Protozoa, rotifers, copepod nauplii, zooplankton, phytoplankton
Fry (24-30)	Cladocerans, copepods, microzoobenthos, detritus, phytoplankton, microalgae, plant fragments	Mainly phytoplankton	Zooplankton and phytoplankton
Fingerlings (up to 100)	Detritus and plants as the pharyngeal teeth develop	Phytoplankton	Mainly zooplankton, some phytoplankton

Table 8.5: Intake of Food Organisms by Common Carp Postlarvae/Fry (Modified from Bryant and Marry 1980)

Stage (Days from First Feeding)	*Food Organism*	*Daily Intake (% Body Weight)*	*Temperature (°C)*
Postlarvae/fry (up to 5)	*Artemia nauplii*	200-500	24
Postlarvae/fry (up to 10)	*Artemia nauplii*	100-120	24
Postlarvae/fry (up to 10)	*Moina* spp.	200-300	23-25
Postlarvae	*Daphnia carinata*	200-300	18-23

Feeding Regiment for Carp Postlarvae and Fry

Hatchlings and fry can be reared in hatchery tanks where their food can be provided in adequate quantities and where their development and state of health can be closely watched. This means that the tanks must be indoors or at least in a roofed area well protected from the weather. Advanced fry, at about 10 days old, can then be transferred to nursery ponds for growout to fingerling size.

Carp hatchlings have a short non-feeding period during which the yolksac is largely, although not completely, absorbed. This can vary from one to five depending on the temperature and the species. Tanks containing hatclings should be kept very clean, removing all dead eggs, egg shells and dead hatchlings. This is to prevent the buildup of harmful bacteria, fungi and protozoa. When the hatchling's mouth opens, are taken in to fill the swim bladder. Active feeding then commences, although the yolksac is still not completely absorbed and continues to supply energy. The recommended maximum density for first-feeding postlarvae is 100,000/m^3.

In recent years, fiberglass, high density polyethylene, polypropylene and other plastic tanks have become widely accepted as essential elements in a high quality hatchery. The details regarding size, material and design vary according to the needs of different hatcheries, but the use of cylindroconical tanks is now becoming widespread. These can be used for hatchlings, fry and for the rearing of live food organisms. There advantages over tanks of other shapes are that they are very easy to keep clean, easy to move around (therefore, giving greater flexibility to hatchery operations than systems set in concrete), easy to fill, empty and observe. Figure 8.1 gives designs for cylindroconical tanks of three capacities : 0.5, 1.0 and 4.0 m^3. The two smaller size are particularly useful for mass rearing of the food organisms such as rotifers and *Artemia* nauplii. Figure 8.2 gives as outflow filter design.

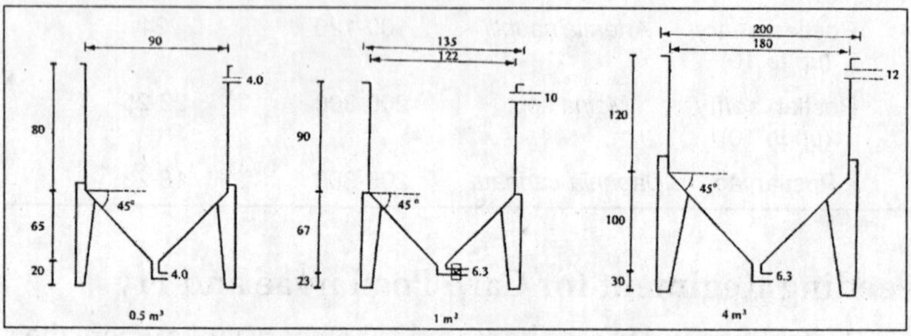

Figure 8.1: Dimensions for Cylindroconical Tanks; All have three legs. Units are centimeters.

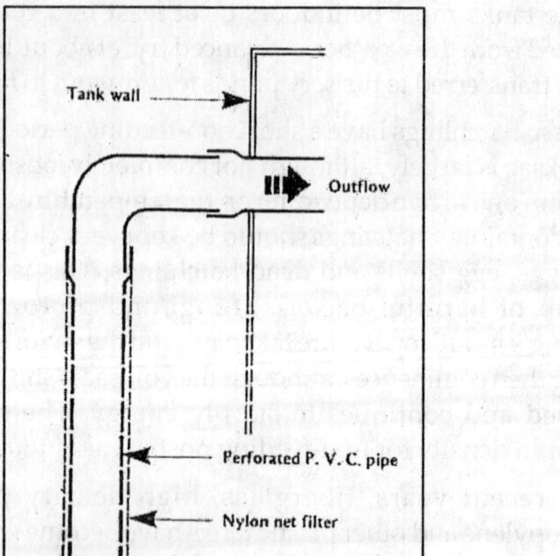

Figure 8.2: Outflow Filter for Cylindroconical Tanks.
Both figures. Dr. Jacques fuchs of Centre Oceanologique Du Pacifique.

Starter Feeds for Hatchlings and Fry in Tanks

Carp fry will accept living and non living food items. The most widely used starter feed are dense cultures of rotifers or brine shrimp

(*Artemia*) nauplii and/ or the so-called microencapsulated egg diet. Theoretically, are one of these could supply the required for the first few days of rearing, but reliance on microencapsulated egg diet alone is not a good idea. It helps to enlarge the gut and make it function but it is not a complete feed. The best carp-rearing results have been obtained by using rotifers as a starter feed and progressing through *Artemia* nauplii to larger crustaceans such as the caldocerans, *Monia, Daphnia* and *Bosmina* (Table 8.6). The most important factor for starter feeds is size. First feeding is usually on particles in the size range 50-100 μm. Artificial carp starter feeds are also now becoming available, for example, Ewos C10 "larvstart" from Sweden. This is a 50-60 per cent protein complete feed, including a vitamin premix and is available in different particle sizes.

Techniques for supplying starter feeds are considered below under the following heading :

Rotifers, Artemia and microencapsulated egg diet.

Table 8.6: Recommended Size/Density Relationships for Early Rearing of Common and Chinese Carp Fry on Live Foods[a]

Postlarvae/Fry Size (mm)	*Food Organisms*	*Food Size (μm)*	*Food Density/ Liter*
6 – 7	Rotifers	120-180	2,000 – 5,000
7 – 9	*Artemia nauplii*	250-600	1,000-1,500
9	Cladoceran (Moina)	1,000	As required

a: After Rothbard (1982).

Rotifers

Rotifers are very easy to culture. In brackishwater and marine larval fish rearing, pure cultures of the rotifer *Brachionus plicatilis* have become the main starter feed used for species with mouths too small to accept *Artemia* nauplii. For carp culture, however, rotifers culture is usually achieved by producing blooms of natural plankton in tanks fertilized with organic manures. *Brachionus* species, for example *B. calyciflorus*, are particularly good feed organisms as the come within the 100-150 μm size range. The brackishwater species *B. pilcatilis* can be used to feed carp postlarvae in freshwater. Rotifers taken from mass culture at 10 ppt salinity survive at least 24 hr in freshwater (Lubzens *et al.*, 1984). Larger carnivorous genera from 200-600 μm cannot be eaten by carps as starter feeds.

The culture of mixed rotifers in green water conditions usually provides an excellent starter feed for all cyprinid early fry. Such cultures are usually also rich in large Protozoa, particularly ciliates, which are additional food items. Mixed plankton ensures that the rotifers have a varied diet. Essential fatty acid deficiencies in *Brachionus* as a food for marine fish larvae have been recorded and are caused by reliance on the use of pure cultures of fatty-acid deficient phytoplankton species to feed the rotifers. *Isochrysis galbana* is particularly good alga for culturing *Brachionus plicatillis* of high food value for marine fish larvae. *Dunaliella tertiolecta*, by contrast, is deficient in long chain polyunsaturated fatty acids (PUFA). If pure culture of a phytoplankton/rotifer species combination is envisaged for carp rearing purpose, then the possibility of such deficiencies should be borne in mind.

Artemia nauplii

For hatchlings/fry of 7-9 mm, *Artemia nauplii* can provided a complete feed. In fact, carps, particularly common carp seem much less sensitive to difference in the quality/origin of *Artemia* cysts than most cultured backishwater and marine fish Tables 8.7 to 8.9 gives some results from a recent study on *Artemia* cysts from various sources used to rear common carp fry. Artemia is a saline water organism, hence, the need to evaluate its survival time in freshwater conditions. The tables show that while survival of common carp fry is good on *Artemia* from all sources, a bioeconomical evaluation based on price, hatching rate, naupliar size and nutritional quality does permit source to be ranked in order of preference. However, sources of *Artemia* cysts are continually changing, batches vary in quality and prices also vary. Therefore, the data in these tables are to be taken as a example of the variation between source and not as a recommendation for buyers. For up-to-date information, hatchery operators are advised to contact the *Artemia* Reference Centre, State University of Ghen, J.Plateaustraat 22, B 900 Ghent, Belgium. Reference cysts of known characteristics (hatching success, biometrics, essential fatty acid content and pesticide content) are available from this Center for fundamental or applied research work. They are known as Reference *Artemia* Cysts (RAC). They are small-sized contain good source of PUFA and low chlorinated hydrocarbon levels and have been tested with a wide range of fish and crustacean larvae.

Table 3.7: Characteristics of *Artemia* Cysts Used to Rear Common Carp Postlarvae

ET50 = median effective time for 50 per cent reduced swimming activity of nauplii transferred to freshwater (nauplii only moving appendages on the container bottom): LT50 = median lethal time for nauplii transferred to freshwater (modified from Vanhaecke and Sorgeloos 1983).

Sources of Cysts	*Hatching Output (mg nauplii/g of cysts)*	*Naupliar Harvesting Time (Hours after hatching)*	*Naupliar Dry Weight (µg)*	*ET50*	*LT50*	*Type of Artemia Strain B: bisexual P: parthenogenetic*
				Hours		
San Francisco Bay, Ca-USA batch no 288-2596	435.5	24	1.63	6.5	26	B
San Pablo Bay, CA-USA batch no 1628	497.7	24	1.92	5.9	28	B
Macau, Brazil batch no 871172	529.0	24	1.74	7.0	27	B
Great Salt Lake Ut-USA batch no 185	467.0	24	2.43	6.8	26	B
Shark bay Australia batch no 114	537.5	29	2.47	5.9	37	P
Chapin Lake, Canada harvest 1978	400.4	27	2.04	4.2	16	B
Lavalduc, France harvest 1979	561.8	31	3.08	8.0	38	P
Tientsin, People's Republic of China harvest 1978	400.5	29	3.09	6.8	33	P
Margherita di Savola, italy harvest 1977	458.2	29	3.33	7.6	36	P

Table 8.8: Survival and Weight of Common Carp Hatchlings/Fry Fed *Artemia nauplii* from Various Sources Data suffixed by the same latter are not significantly different ($p < 0.05$)[a]

Sources of Cysts (see Table 8.7 for details)	*Survival (per cent)*	*Mean Individual Weight After 7 Days (mg)*		*Mean Individual Weight After 14 Days (mg)*	
		$\overline{X}$	*S*	$\overline{X}$	*S*
Margherita di Savoia	94.3	31.9	2.0	188.4	8.0
Tientsin	96.7	32.2	2.1	184.9	5.3
Shark Bay	95.7	31.8	2.1	180.4	5.2
Lavalduc	95.2	31.0	2.8	179.5	5.5
Macau	95.7	30.3	2.2	173.1	2.9
Great Salt Lake	93.3	29.7	1.5	170.7	2.3
San Pablo Bay	93.3	29.9	2.3	169.5	5.1
San Francisco Bay	93.3	29.9	3.6	166.4	12.3
Chaplin Lake	95.2	24.6	0.9	143.4	0.2

a: After Vanhaecke and Sorgeloos (1983).

Table 8.9: Bioeconomical Evaluation of the Use of Specific *Artemia* Cyst Batches from Various Geographical Origins for Carp Hatchlings/Fry

Sources of Cysts (see Table 8.7 for details)	*Quantity of Cysts Needed for One Car Larvae (in mg)*		*Quantity of Cysts Needed for the Production of 1g Carp Biomass (in g)*	
	In 1 week	*In 2 weeks*	*In 1 week*	*In 2 weeks*
Lavalduc	24.5	174.0	0.79	0.98
Shark Bay	25.7	181.9	0.81	1.01
Macau	26.1	184.8	0.86	1.08
Margherita di Savola	30.2	217.7	0.95	1.16
San Pablo Bay	27.8	196.4	0.96	1.17
Great Salt Bay	29.6	209.5	1.00	1.24
San Francisco Bay	31.7	224.5	1.06	1.36
Tientsin	34.5	244.1	1.07	1.32
Chaplin Lake	34.5	244.1	1.40	1.72

a: After Vanhaecke and Sorgeloos (1983).

Artemia cysts can be simply incubated in seawater (natural or artificial) to provide nauplii, for example, 3 g cysts/liter of seawater (around 35 per cent salinity). The hatched nauplii must be separated from the hatching debris which may otherwise block the digestive system of the fish hatchlings if ingested. However, recently a method known as decapsulation has gained popularity since it make naupliar hatching easier and increase hatching success. Decapsulated cysts can also be used as food before they hatch.

The following method is used for decapsulation of 10g of cysts (details as published by Rothbard 1982):

Step	*Procedure*
1	Imerse the *Artemia* cysts in 2 liters of strongly aerated seawater. This rehydrates the cysts.
2	Pour through a 150-200 µm net and wash several times with seawater.
3	Make a solution of 292.5 ml seawater plus 5 g of active sodium hypochlorite.
4	Pour the cyst into the solution and stir in slowly and carefully 7.5 ml of 40 per cent sodium hydroxide solution. The dilution of this strong alkali generates heat so addition must be very slow and with stirring. The temperature must not rise above 40°C since this will kill the *Artemia*. Check with a thermometer and add small piece of ice if necessary. Stir constantly.
5	The suspension of cysts will change colour within 7-15 minutes as decapsulation occurs. At this point, rinse the cysts again in seawater using the fine mesh net and continue rinsing until the small of hypochlorite has disappeared.
6	Place the decapsulated cysts in 200 ml of seawater and 0.5 ml of 1 per cent sodium sulphite with stirring. This neutralizes any remaining hypochlorite. Stir for several minutes.
7	Let the decapsulated cysts sink to the bottom of the vessel and pour off any floating debris.
8	Incubate decapsulated cysts in seawater for hatching in the normal manner.

Microencapsulated Egg Diet

Whole chicken egg is one of the most well-balanced food and has long been used as a yardstick in nutritional science. Table 8.10 shows its composition.

Table 8.10: The nutrient Composition of Chicken Egg

	Whole Egg	*Egg White*	*Egg Yolk*
Protein (per cent)	48.8	76.9	32.8
Fat (per cent)	43.2	-	62.2
Gross energy (kca/kg)	5,830	3.070	6.910
Metabolizable energy (ME) (kcal/kg)	4,810	2.533	5.700
ME : protein ratio	9.8	3.3	17.3
Calcium (per cent)	0.2063	0.0427	0.2653
Phosphorus (per cent)	0.873	0.282	1.020
Amino Acids (per cent)			
Arginine	2.968	4.179	2.369
Cystine	0.887	1.282	0.526
Isoleucine	2.734	4.307	0.896
Leucine	2.063	6.273	2.790
Lysine	3.047	4.427	2.369
Methionine	1.563	2.700	1.663
Phenylalanine	2.500	4.427	1.316
Threonine	2.500	3.692	1.843
Tryptophan	0.887	1.350	0.577
Tyrosine	1.952	3.076	1.316
Valine	3.674	5.025	2.263

After Chow (1980).

Hard-boiled eggs have been used as supplemental feed in fish rearing for a long time. For microencapsulated egg diet, however, the principle is to mix whole egg homogenate with boiling water, whereupon a suspension of miniature hard-boiled egg is formed–each yolk particle being encapsulated in a coating of denatured aluminous protein. The procedures for this are as follows (after Chow 1980; Rothbard 1982) :

Step	*Procedure*
1	Break egg into a heat-resistant container.
2	Beat egg vigorously on homogenize with a mechanical blender.
3	Pour rapidly (approximately 150 ml for each egg boiling water into homogenate while stirring constantly. A fine, opalescent suspension is obtained.
4	Make up to desired volume with cold water. A 50 g egg contains about 12 g dry matter.
5	Feed by the spoonful of scoopful directly to fish. The feed may also be applied as a spray using a "kynapsack" type sprayer.
6	Store unused feed in a sealed container at 0.4°C.

As stated above, this feed is best regarded as supplemental to a live food diet. The best principle for feeding carp fry is to provide the them with a abundance of food items so that growth can be rapid, always taking care, however, now to foul the tank with decomposition of uneaten food. Uneaten egg diet is particularly dangerous since its decomposition generates hydrogen sulphide (H_2S).

The microencapsulated egg diet can also be modified as follows for supplemental feeding to advanced fry which require larger food items :

Step	*Procedure*
1	As above
2	As above
3	Pour homogenate directly into boiling water. Amount of stirring depend on feed particle size desired. Initially, strands of cooked egg are formed. Dispersal of egg yolk remains the same. Yolk is still encased in denatured (cooked) albumin.
4	As above
5	Feed resuspended material directly to fish. Do not apply with sprayer as this will result it feed particles of undesirably small size for large fry.
6	As above.

Alternatively

Step	*Procedure*
1	Add 50 ml water and process as above.
2	Steam homogenate in same container or after transfer to appropriate dish or bowl. A custard made up of the solidified egg emulsion is formed.
4	Feeds custard directly to fish after breaking it up into desired size.
5	Store unused custard in tight container under refrigeration.

Cladocerans ('Water Fleas') as Feed for Carp Fry Beyond Starter Diets

Cladocerans are used for feeding fry large enough to accept larger food items than rotifers and *Artemia* nauplii. Caldocerans are cultured in small pond, tanks or plastic pools in a similar manner to that described above for rotifers. The most common genera cultured are *Bosmina, Moina and Daphnia. Moina* is particularly important as a fry and fingerling food organism in warm water aquaculture in Asia. Culture methods involving soil and animal manure addition to pond water have been described for *Moina macrocopa* in the Philippines since 1921. However, soil is not really necessary.

Essentially, the methods for all cladoceran culture rely on animal manure. For example, *Moina micrura* in India is cultured in plastic pools by an initial fertilization with fresh cowdung and groundnut cake at a rate of 250-350 and 500 ppm, respectively, following by subsequent application of half of these quantities every 4 days. The cultures are seeds with 2-5 ml *Moina* collected from local sources. This system can yield a harvest of 30-70 ml of *Moina* and other plankton, every two days from a 300 liter pool for about a 12 day period. These cultures are not pure cultures of the desired organism. They are best termed semi-pure. In order to get a higher percentage of the desired organism, the original seeding should be done very carefully, with care not to introduce other species and all water should be tap water or pond water filtered through fine bolting silk. The *Moina* or other caldocerans can be collected and fed directly to fry in rearing tanks or used to inoculate outside fertilized ponds. The rate of inoculation for outside pond is 30-50 ml *Moina*/ha.

Feeding Advanced Fry Fingerling in Nursery Ponds

The provision of foods for advanced fry and fingerling of cultured carps is usually achieved by fertilization of nursery ponds to stimulate the production of plankton, benthic microalgai and micro- and meiofauna with supplemental feeding of powdered feeds such as rice bran and peanut cake. Here again, feeding practices have been developed empirically and not on any critical analysis of nutritional requirements. Ponds fertilization with organic manures and inorganic fertilizers is discussed in detail in Chapter 7.

Trends in Artificial Feeds for Postlarvae, Fry and Fingerlings

In the future, feed manufactures will continue to improve carp growout and broodstock feeds, striving for least cost formulation and using cheaper alternatives to fish meal and animal proteins. The most important developments, however, are likely to occur in hatchling fry and fingerling nutrition as complete feeds are produced to take the place of live food organisms. Research on the nutrition of carp fry and fingerlings is continually being carried out and the results are being used to formulate artificial feeds. The provision of complete artificial feeds at realistic prices would constitute a considerable advance over the use of live food organisms and location-specific fertilization/feeding practices.

For grass carp fry, high protein feeds in the 41-53 per cent protein range seem to be required and other species are probably similar. Experimental rearing of grass carp fry using yeast (*Candida lipolytica*) as a protein-rich feed has been very successful. Table 8.11 gives a comparison of the composition of the yeast and fish-meal.

An experiment in which grass carp postlarvae were fed on yeast supplemented with a 4-5 per cent vitamin/mineral premix and 5.5 per cent meat extract produced excellent result (Figure 8.3). A trial with common carp has shown that the yeast feed out performs a conventional *Artemia* nauplii plus boiled egg yolk diet (Figure 8.4) (Appelbaum and Dor 1978). The total amount of dry yeast feed required here was about 250g for 50,000 hatchlings is a 100 × 100 × 60 cm plastic tank. Feed was scattered over the water surface 6-7 times a day : 15g/day for the first few days with grain size not larger than 250 μm increasing to 25g/day of grains up to 500 μm from the third day and up to 40g/day of 750 μm grains towards to end of the

Table 8.11: Analysis of the Artificial Food-Years *Candida lipolytica* in Comparison to Fishmeal

	Yeast	Fishmeal		Yeast	Fishmeal
Dry matter–per cent	94.4	89.1	Amino Acids		
Ash–per cent	6.3	24.7	Histidine	2.1	2.4
Organic substance (per cent of dry matter)	93.7	75.3	Leucine	7.2	7.3
Crude protein–per cent	64.4	71.8	Isoleucine	4.8	5.7
Fat–per cent	6.3	2.4	Lysine	7.5	9.0
			Methionine	1.8	2.7
Amino Acids:			Phenylalanine	4.4	4.2
Alanine	7.4	–	Proline	4.1	–
Arginine	5.2	5.8	Serine	4.5	–
Asparagine	9.0	–	Threonine	4.9	3.9
Cysteine	1.1	0.9	Tyrosine	3.4	3.0
Glutamine	12.7	12.5	Valine	5.7	5.1
Glycine	4.9	6.1	Tryptophan	1.3	1.0

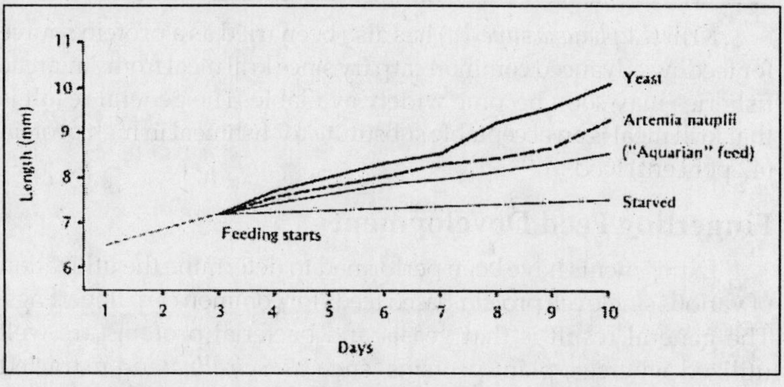

Figure 8.3: Growth of Grass Carp Postlarvae/Fry Fed on Different Diets. Water temperature 24° C; stocking density approximately 125 individuals/liter; data represent averages of 30 measurements each (from Appelbaum and Uland 1978).

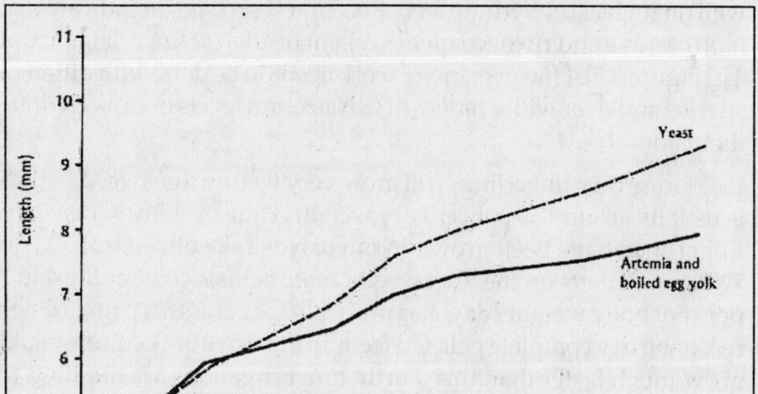

Figure 8.4: Growth of Common Carp Postlarvae/Fry Fed with Dry Yeast Feed or Artemia Plus Boiled Egg Yolk (From Appelbaum and Dor 1978)

trial. Such trials require repetition and scaling-up for commercial use. It is very important to remember that yeasts, like all protein-rich feeds, can cause tank fouling, depletion of oxygen water exchange or cylindroconical and specialized carp rearing tanks are essential.

Krill (*Euphausia superba*) has also been tried as a protein source for feeding advanced common carp fry since krill meal from Antarctic fisheries may soon become widely available. The general result is that krill meal is an acceptable substitute for fishmeal in high protein (42 per cent) feeds.

Fingerling Feed Development

Experiments have been performed to determine the utilization of various single cell protein-based feeds for common carp fingerlings. The general result is that yeasts and bacterial proteins are well utilized whereas, plant proteins, such as *Spirulina* and extracted soybean proteins concentrate, are not. It is possible the future fingerling feeds may be based on petroyeasts or methanophilic bacteria.

It has also been shown that common carp fingerlings will do well on high carbohydrate feed. Recent research result indicate that high cassava and rice feed give excellent results for fingerlings about 4 g (Figure 8.5). However, more work needs to be done with different species and over wider ranges of fish size and levels of carbohydrate inclusion.

Grass carp fingerlings will grow very well on duckweed (which is usually around 30 per cent dry weight crude protein). Grass carp fingerlings have been grown in intensive tank culture from 2.7 to 72.2 g is 88 days on the duckweed *Lemna minima* consuming 4 to 7 per cent body weight/day. It is probably that grass carp fingerlings receiving dry complete pelleted feeds in intensive tank culture could grow much faster than this. Further investigations are required to exploit the growth potential of this species.

The planktivorous carps are probably best left to exploit their preferred natural feeding niches as fingerlings and through growout, rather than attempting to develop feeds for intensive culture.

Feed Analysis and Formulation

The following information is from Jauncey and Ross (1982). "A guide to tilapia feeds and feedings", since their recommendations for analysis, quality control and formulation principles can also be applied to carp feeds. The feeds in question are best consideration complete rather than supplemental feeds.

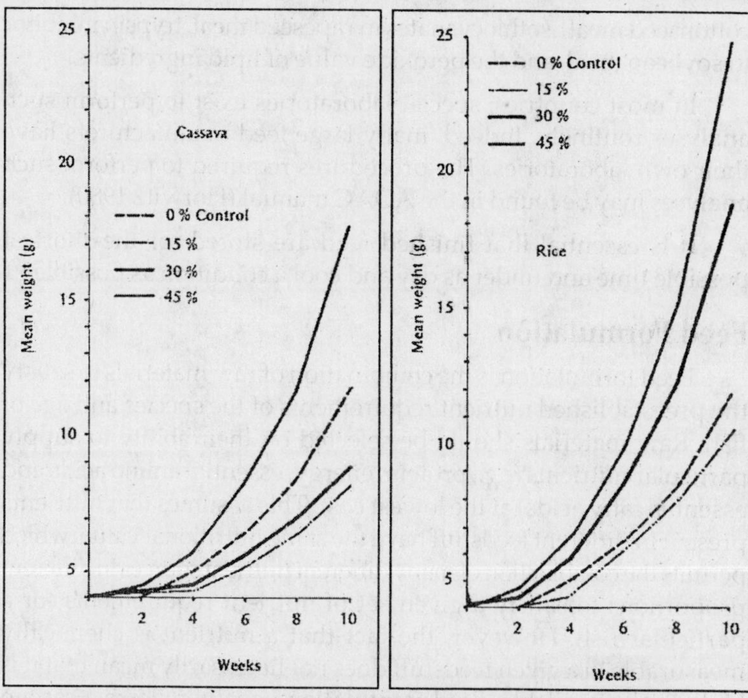

Figure 8.5: Effects of Different Levels of Dietary Cassava and Rice on Weight Gain of Mirror Carp Fingerlings (From Ufodike and Matty 1983)

Feed Analysis and Quality Control

Much information can be gained about nutritional composition of any given feedstuff by consulting tables of average analyses of the material in question. Useful references to consult for this are: "Tropical Feeds" (Gohi 1975) and "Nutrient requirements of warm water fishes" (NRC 1977, 1983). However, it should be appreciated that the composition of any given feedstuff may vary enormously from batch to batch. It is desirable, therefore, to have batches of ingredients and the finished feed sampled and analyzed, at the very least proximate analysis for protein, lipid, fiber, ash and carbohydrate. Ideally, analyses should also be carried out for essential amino acids, essential fatty acids and minerals, although this is often not practicable. The levels of suspected toxins should also be examined in susceptible feedstuff, *e.g.*, aflatoxins in oilseed meals, gossypol in cottonseed meal, isothiocyanates in rapeseed meal, trypsin inhibitor in soybean meal, and the peroxide value of lipid ingredients.

In most countries, special laboratories exist to perform such analyses routinely. Indeed, many large feed manufacturers have their own laboratories. The procedures required to perform such analyses may be found in the AOAC manual (Horwitz 1980).

It is essential that finished feed are stored for the shortest possible time and under as dry and cool a condition as possible.

Feed Formulation

Feed formulation is the combination of raw materials to satisfy the preestablished nutrient requirements of the species and age of fish. Raw materials should be selected on their ability to supply particular nutrients (*e.g.*, protein, energy, essential amino acids and essential fatty acids) at the lowest cost. This assumes that nutrients present in different feedstuff have the same nutritional value, which permits the combination of many different nutrient sources in different proportions to satisfy a given set of nutrient requirements for a particular fish. However, the fact that a nutrient is chemically measurable in a given feedstuff does not necessarily mean that it is biologically available. Feed formulation usually follows a certain sequence of virtually trial and error steps:

Step	*Procedure*
1	Balance the crude protein level.
2	Check and balance the energy content.
3	Check the levels of essential amino acids and essential fatty acids and if necessary, return the amino acid requirements are satisfied, and return to step 2 to adjust the lipid sources until the essential fatty acid requirements are satisfied.

Before proceeding with the above steps, however, it is usually advisable to use least-cost or best-buy calculations to compare the cost of supply a particular nutrient from different feedstuffs. For example :

If cottonseed meal costs Rs. 18.50/kg and contains 54 per cent protein :

$$\text{Cost/kg protein} = \frac{18.50}{0.54} = \text{Rs. } 34.26/\text{kg}$$

If groundnut Meal cost Rs. 14.10/g and contains 44 per cent protein :

$$\text{Cost/kg protein} = \frac{0.3071}{0.44} = \text{Rs. } 32.05/\text{kg}$$

Thus, although the groundnut meal in question has a lower protein content, it is still a better buy as it costs less per kg of protein supplied.

If rice polishings contain 2,700 kcal DE (digestible energy)/kg and cost Rs. 7.38:

$$\text{Cost/kcal} = \frac{7.38}{2{,}700} = \text{Rs. } 0.003/\text{kcal}$$

If maize-hominy feeds contain 4,300 kcal DE/kg and cost Rs. 9.83:

$$\text{Cost/kcal} = \frac{9.83}{4{,}300} = \text{Rs. } 0.002/\text{kcal}$$

Thus, as an energy source the maize-hominy feed is cheaper per kcal of energy supplied than the rice polishing eventhough it is more expensive.

Similar calculation can be performed to find the least-cost feedstuff for supplying any given nutrient.

Balance Crude Protein Level

Protein is the most expensive component of carp feed and is the first nutrient considered in diet formulation. Protein sources are first selected from those available on their cost per unit of protein supplied and their amino acid profiles. Supposing it is desired to formulation a 30 per cent protein carp feed from cottonseed meal (54 per cent protein) and maize germ meal (20 per cent protein), a square is constructed thus :

Cotton seed meal 54% — 30 – 20 = 10

Desired feed protein level
30 per cent

Maize germ meal 20% — 54 – 30 = 24

The protein level of the feed is subtracted from that of each of the feedstuffs in turn and the answer is placed at the opposite corner to the feedstuff ignoring positive of negative signs. The two figures on the right hand side of the square are then added together (10 + 24 = 34). To obtain the 30 per cent protein carp feed we need :

$$\text{Cotton seed meal } \frac{10}{34} \times 100 = 29.41\%$$

$$\text{Maize germ meal } \frac{24}{34} \times 100 = 70.59\%$$

Thus, to make 100 kg of the 30 per cent protein carp feed we need 29.41 kg of cotton seed meal and 70.59 kg of maize germ meal. If there are more than two protein sources to be taken into consideration then this is achieved by grouping the feedstuffs, on the basis of their protein content, into protein supplement (> 30 per cent protein) and basal feed (< 30 per cent protein).

For example, the 30 per cent protein carp feed could be formulated from rice polishing (14 per cent protein) and soybean meal (48 per cent protein) in addition to the cottonseed and maize germ meal used above. The protein supplement is an equal mixture of soybean meal and cottonseed meal and has a protein content of $\frac{54 + 48}{2} = 51\%$. The basal feed is an equal mixture of maize germ meal and rice polishing and has a protein content of $\frac{20 + 14}{22} = 17\%$.

To obtain the desired 30 per cent protein level, a square is constructed thus :

Protein supplement 51 per cent — 13

30

Basal Feed 17 per cent — 21

We thus require $\frac{13}{34} \times 100 = 38.24\%$ protein supplement and

$\frac{21}{34}$ $34 = 61.76\%$ basal feed

One hundred kg of feed would, therefore, contain 19.12 kg each of soybean meal and cottonseed mean and 30.88 kg each of maize germ meal and rice polishings.

Exactly the same procedure can be used with digestible energy values to arrive at the final desired digestible energy content of the feed.

Complete Feed Formulation

The procedure used above account for 100 per cent of the feed and leaves no room for supplementary lipid, binders, vitamins and minerals. If we assume that all the supplements combined account for 10 per cent of the feed then the calculation is adjusted by multiplying the figures derived from the square by 90 instead of 100.

Step	*Procedure*
1	Use least cost analysis to select protein and energy sources.
2	Ensure that these source will provide the desired levels of essential amino acids (EAA) and essential fatty acids (EFA).
3	Remember that a better nutrient balance is likely to be achieved by using several feedstuffs in combination.
4	Balance crude protein level.
5	Balance digestible energy level.
6	Calculate the levels of EAA and EFA in the finished feed (if possible) and if these do not satisfy the requirement of the carp, repeat steps 4 and 5.

The best way to undertake step 6 is to produce a feed formulation table (see Table 8.12 to 8.13)

This method of diet formulation may appear laborious at first, but practice and experience allow the feed formulator to "Guesstimate" the required levels and types of feedstuff and then use the calculation to check the nutrient requirements are satisfied, drastically shortening the whole process.

Linear Programming

The best method of diet formulation is to use an appropriate computer program. The computer stores the complete analysis and cost of the available feed stuffs and the required nutrient profile of the feed. It will perform calculations to prepare a least-cost formulation for any required feed virtually instantaneously. Such computer software can be bought off the shelf and the hardware systems are readily available. However, such systems are only economically viable if the quantities of feed being produced are very large.

Processing Ingredients

Feedstuffs are processed in order to increase digestibility and pelletability and to inactivate specific antinutritional factors such as growth inhibitors or toxins. The simplest form of feedstuff pre-treatment is milling which is applied to materials with low moisture content. This reduces particle size to improve pelletability and digestibility. Pellets containing feedstuffs milled to a uniformly small particle size are also much more water stable. There are many different type of milling equipment available and their use depend, to some extent, on the nature of the feedstuff, particularly the lipid and fiber contents. Ball mills, hammer mills and cutter mills are the three main types employed.

Heal treatment is also often used for drying feedstuffs, destroyed antinutritional factors (such as the trypsin inhibitor in soybean) or in the extraction of oil from oilseeds. Wet heat (steam) improves the nutritional value of the any feedstuff or feed containing large quantities of starch by gelatinization (conversion to destrins and oligosaccharides) which result in the carbohydrates becoming more digestible. Gelatinization of the starch also greatly improves its binding properties. There are many different kinds of pelleting

Table 8.12: Worksheet for Diet Formulation with Respect to Major Nutrients and Cost (After Jauncey and Ross 1982)

Ingredient	% in	Cost/100 kg		% Protein		% Lipid		% CHO[a]		% Fiber	
	Feed	Ing[b]	Feed	Ing	Feed	Ing	Feed	Ing	Feed	Ing	Feed
Totals	100	–	–	–	–	–	–	–	–	–	–
Req	100	–	–	–	–	–	–	–	–	–	–

a: CHO: Digestible carbohydrate.

b: Ing: ingredient (feedstuff).

c: Req: Requirement.

Table 8.13: Worksheet for Diet Formulation with Respect to Essential Amino Acids (After Jauncey and Ross 1982)

Ingredient	% in	% Arga		% Hist		% Isoleu		% Leu		% Lys		% Math		% Phen		% Thr		% Try		% Val	
	Feed	Ing	Feed	Ing	Feed	Ing	Feed	Ing	Feed.	Ing	Feed	Ing	Feed	Ing	Feed	Ing	Feed	Ing	Feed	Ing	Feed
Totals	100	–	–	–	–	–	–	–	–	–	–	–	–	–	–	–	–	–	–	–	–
Req	100	–	–	–	–	–	–	–	–	–	–	–	–	–	–	–	–	–	–	–	–

Abbreviations for the ten essential amino acids.

equipment available. The simplest (for processing small quantities) are large commercial food mixers which can be used to produce a wet-extruded diet which is subsequently dried. On a larger scale, machines exist for producing dry extruded pellets-these simply force the feed mix through a die under pressure. Their disadvantage is that the pellets produced are extremely hard.

Pellets can produce in the required sizes from about 2 mm in diameter upwards and, ideally, the length of a pellet should be 2-3 times its diameter. Smaller feed size for fry and fingerlings can be produced by crumbling larger-sized pellets and sieving them to the appropriate size although special machines are available for the production of crumbs.

Table 8.14: Ingredient and Percentage Composition of a Supplemental Feed for Common Carp Broodstock

Fish and/or meat meal (blood meal)	22
Wheat, sorghum, com (or similar grain) And some wheat or rice bran	30
Extracted soybean meal	17
lupin, pea or other legumes (partly extracted oilcake)	10
Maize	14
Flour of alfalfa (Lucerne) or clover	5
Yeast	0.5
Vitamine premix	0.5
Mineral premix	0.5
Stabilizer	0.5

Some Aspects of Feeds for Broodstock

Sufficient natural food can be produced in a well-managed and adequately fertilized pond large enough to provide qualitative and quantitative nutrition for the stock. A 20 to 40 m^2 pond surface can produce the necessary food for one female. Utilization of the protein-rich natural food will be improved if a supplement of 1 per cent of the body weight of each breeder is given daily in the form of a carbohydrate rich grain feed. If the pond surface is less the 20 to 40m^2 per female, a complementary protein-rich feed has to be given daily in a ratio of 2 to 2.5 per cent of the body weight. Other suitable supplementary feeds are the protein–rich grain such as soybean, beans, peas, lupins, good brans, oilcakes, etc. Females need a diet rich in animal protein, vitamins and minerals for the development of the oocytes in the ovary. Males do not need such rich feed as the females. The protein content of their supplementary feed can be reduced to about 15 to 20 per cent of crude protein. However, natural food is the most important part of the diet of all the broodstock. Table 8.14 gives a typical feed mix for a supplemental feed for common carp broodstock, containing 29-30 per cent crude protein. If such a feed mixture is not readily available, the feed given to laying hens (in a pelleted form or in a dough but not in dry meal form) can be used. It is given together with about 10 per cent of fish, meal, blood meal, or fresh blood, or with minced part of other animals, such as fish, frogs, tadpoles, insects (beetles, locusts), earthworms or silkworm pupae. More information on broodstock nutrition is tabulated in Table 8.14.

Chapter 9

Guidelines for the Construction of Fish Seed Farms/Hatcheries Adopting Improved Designs

Introduction

Due to heavy exploitation of riverine fishery resources and growing pollution of rivers availability of carp seed from this traditional source is declining at a fast rate. Moreover, riverine fish seed collection by traditional practice is now limited and hence dependence on riverine fish seed source is not at all desirable and dependable.

Quality fish seed is the basic input for successful carp culture. The central Inland Fisheries Research Institute (CIFRI) of Indian Council of Agricultural Research (ICAR) have successfully developed the techniques of induced breeding the Indian and exotic carps in confined waters through the administration of pituitary gland (P.G) extract but the techniques have not been adopted so far for the production of quality seed on commercial scale in states due to non-availability of fish seed farms/hatcheries providing the required facilities such as temperature controlled hatcheries, improved designs for dewatering of ponds for renovation and efficient operation of these farms. The technology of induced breeding has also not been adopted in the different States of India due to lack of technical expertise for the construction of technically sound hatcheries and their operation.

The eggs are incubated normally in a cloth chamber (enclosures known as hapa) in the pond water. The eggs while developing usually are subjected to heavy mortality due to inadequate dissolved oxygen in water wide and rapid fluctuation of water temperature and also due to presence of predators such as crabs, snakes, common carp and other fishes. Because of these constraints there is very heavy mortality of the developing eggs. To overcome all these Central Inland Fisheries Research Insitute, Barrackpore (Kolkata, West Bengal) and Central Institute of Fisheries Education (CIFE, Mumbai) of ICAR have developed small indoor hatcheries, using jar or net hapas for hatching and PVC tubs for rearing hatchlings/fry.

During 6th plan Fish seed Farms/hatcheries of size 10-25 ha (Production capacity 10-27 million fingerlings) have been constructed under the World Bank Assitance Programme in Madhya Pradesh, Orissa, West Bengal, Bihar and Uttar Pradesh. While operation of these Fish Seed Farms/ hatcheries built on World Bank design. Difficulties have been experienced in filling/draining out of pond water, harvesting of fingerlings through the central channel and adoption of stripping techniques for collection and fertilization of eggs. These difficulties had been overcome in the present improved designs, where separate inlets for water filling and outlets for draining out of ponds through the harvesting boxes have been provided for independent operation of each pond. These improved designs are being now adopted for the construction of fish Seed Farms/Hatcheries in the other States under the National Programme for Fish Seed Development.

In the present improved designed given here, either Indoor hatchery building including one or two circular spawning tanks of circular hatchery building may be adopted in this Fish Seed Farm complex. For the large scale production of spawn, circular hatchery has proved to be successful due to its low cost of construction and semi-automatic operation technique requiring no stripping. Due to the impact of adoption of circular hatchery in West Bengal, fish seed production targets in West Bengal were not only achieved but were exceeded considerably.

After that FAO Fisheries Technical Paper No. 168 on the report of FAO Fisheries (Aquaculture) Mission, who visited China 21 April–

12 May, 1976, was circulated by FAO in June, 1977 and several fisheries scientists/technicians from India has visited China during the recent past, the Circular Fish Hatchery design is practiced in China also now been tried and preferred in India due to the reasons that cost of construction and operation of circular hatchery is less as compared to any other design for the same production capacity. Moreover in the operation requiring great skill, is eliminated. Therefore the circular hatchery system is not considered to be much suitable for the production of quality fish seed in India too.

During the 6th plan it was targeted to produce 2,224 million fish seed. By the operation of hatcheries constructed under World Bank programme and other State Govt. hatcheries, 5421 million fish seed was produced during the terminal year of 6th plan. Now it is tragetted to produce 12,000 million fish seed in the country by the terminal year of 7th plan.

The paper on guidelines for the construction of commercial fish seed farms/hatcheries, adopting improved designs, would be helpful for use of extension workers, bank officials, progressive entrepreneurs and fish farmers.

Planning and Designing of Modern Fish Seed Farms/ Hatcheries

The construction of these fish seed farms/commercial hatcheries call for the an uniform design/construction according to State/local conditions and preparation of project documents incorporating the relevant data on suitability of the site selected, the design, plan, estimates, etc. for construction/operation of the hatcheries/seed farms.

The hatchery/fish seed farm complex would include an indoor hatchery unit, earthern ponds, water gates, central channel, tubewells, administrative building, staff residence, storage building, perimeter fencing and internal roads, etc. Each hatchery would provide facilities to produce hatchlings by induced breeding (hypophysation), nursing and rearing them to fry state (approximately 9 and 18 ponds each of 0.55 ha water area respectively for 5 ha and 10 ha. Hatchery). About 25 per cent of this area would be used for brood stock maintenance. The brood stock will be replaced to the extent of $1/3^{rd}$ each year to keep a healthy brood stock for undertaking breeding operations. A provision will also be made for easy communication from hatchery to the main road, electric supply and other facilities. On full operation and development, it is envisaged that 5 million and 10 million fingerlings of 5.0 cm size would be produced by each 5 ha and 10 ha hatchery respectively.

Since Corporations/FFDAs (Fish Farmers Development Agency) have to undertake fish seed production on commercial scale, with this background, it is necessary that due consideration is given by the Corporations/States in respect of site selection and preparation of design and estimates for construction of hatcheries.

Criteria for Site Selection

Selection of hatchery sites will be based on technical and socio-economic criteria, considering the following factors for each site :

1. Availability of land in a contiguous, suitably shaped plot of sufficient size for all facilities (approximately 14 ha, or land for 10 ha hatchery and 8 ha, of land for 5 ha, hatchery.
2. Availability of assured water supplies of adequate quality either surface of ground water.
3. Site surveys and soil testing to determine the suitability of the site for the engineering works envisaged.
4. Soil testing to determine suitability for fish culture.
5. Freedom from frequent deep flooding such that the 10 year return flood is not higher than about 1.85 meters above the average original ground level.
6. Reasonable proximity to the area of demand for fingerlings.
7. Proximity to all weather road connections.
8. Availability of electricity and, if possible, telephone connections; and
9. Avoidance of significant social displacement, particularly in heavily populated areas.
10. Availability of natural drain/nallah nearby the hatchery complex for discharging of waste water from the pond complex.

States in Site Selection

A short list of site should be drawn up and on the basis of existing and easily obtainable information, the sites should be ranked in order of suitability in accordance with the guidelines given under subheading-criteria for site selection.

Then, on outline planning exercise should be carried out on each site. The objective of the exercise is to confirm that a hatchery can be developed on the site in question which is capable of meeting the general criteria and the functional requirements of acceptable capital cost. It is also emphasized that the qualities of the sites selected will have significant influences on capital and operating costs and sites with the most favourable characteristics *i.e.* flat with impervious soils are greatly to be preferred.

Preparation of Plans, Designs and Estimates for Hatchery Construction

The following information are to be incorporated in the plan, design an estimates.

Topographic survey of the site with contours at 50 cm intervals of the site in relation to the data collected.

The flood and drainage conditions.

Detailed survey map showing the soil sampling points, the total number of soil samples and the depth of which samples were collected from the site. A minimum of 6 sample stations for 10 ha and 4 for 5 ha fish seed farm should be used. More stations will be required for variable soil conditions. Disturbed samples should be taken at the interval of 30 cm of depth to a level of 1.5 m below the deepest excavation.

The result of laboratory analysis of the soil samples, indicating the chemical and physical properties of the soil such as soil pH, moisture content elasticity of the soil, grain size, clay content, ISI classification of soil, etc.

Water-table conditions and its seasonal variations.

Details on the evaporation and seepage losses and the total water requirements of initial filling of ponds and for subsequent operation.

Number of tubewells required for the site and details of alternate arrangements for supply of water to the ponds and indoor hatchery system which requires cool water (18-25° C)

Quantum of water which will be available from each tubewell/ other sources during different seasons for initial filling of ponds, after taking into consideration the loss by seepage and evaporation, substantiated by the results on the water table conditions.

The detailed estimate for site preparation, water supply system provision of tube-wells, earthwork, clay soil, bank protection, pitching central canal, site fencing, surfacing of the hatchery roads, water gates, hatchery building, water supply arrangements, administrative buildings, staff and quarters, storage building, fencing and gates, electricity supply and engineering cost could be given. The detailed cost estimated of each hatchery complex could be worked out precisely for a site as per outline rough cost estimates given in Table 9.6 including estimates of stores and equipment given in Table 9.7. The detailed estimates for laying approach roads to connect hatcheries to the main road such as bed formation, pavement construction, bituminous carpeting, drainage work and other engineering works including the cost of the land could be incorporated.

Design for Construction of the Hatchery

Drawings showing the proposed, general arrangement of the hatchery in the proposed site including important cross sectors of the pond water levels in relation to site ground level, flood level and drainage levels, could be given.

Detailed drawings on the water gates, building approach roads, canals, indoor hatchery etc. could also be incorporated alongwith suitable cross sections.

Detailed design for construction of hatchery buildings, administrative building, staff quarters, storage building, fencing and gates, reservoir tanks etc. could also be incorporated into the report.

In order to execute all the civil works and supply of vehicles and equipment, suitable contractors/ tenders have to be predetermined well in time and their pre-qualifications as acceptable could be determined.

Planning

After giving the cost estimates and detailed designs for construction of hatchery in a brief write-up, the proposed stage of site evaluation and development, tendering of work and construction period may be indicated.

CPM chart indicating the programme of implementation including the period of detailed design and tender documentation, approval of estimates, tendering period, report on the tenders, contracting the work and execution of the construction programme including seasonal delays, etc. could also be given in addition to the brief write-up.

All these activities have to be completed within the stipulated period so that the operation of fish seed farms could be started during monsoon.

Other Instruction to be Followed while Preparing a Detailed Project Report

Before describing the site or estimates, plans and design, it is necessary that a brief introduction is given on the project to be taken up in the State under the respective programmes.

The water area to which the proposed hatchery will supply fish seed, number of hatcheries to be constructed with their size and location, likely fish seed to be produced by the individual hatcheries, etc., may find a place in the introduction part.

Functional Requirements of Hatcheries

The main functional requirements of hatcheries and the resulting essential engineering requirements which must be taken into account in the selection of sites and the design of the hatchery layouts are the following :

Ponds must be drained out several times a year for treatment to maintain productivity. This requires that ponds shall be constructed with the bottoms above normal site water table.

It is necessary to be able to maintain any water depth in the pond upto about 2 meters throughout the year and to be able to vary water depth quickly and easily.

This requires :

1. A reliable supply of water throughout the year sufficient to cater for the peak losses due to evaporation, seepage and operational uses of water.
2. Ponds to store water upto about 1.5 m above general ground level. Sites with suitable soils and adequate specifications and workmanship for earth works must be provided to reduce seepage losses to an acceptable minimum.
3. Water distribution and drainage facilities including a central feeder channel, water gates, high capacity portable pumpsets and a drainage channel.

The facilities must have reasonable freedom from flooding. An indoor hatchery is required with in adequate supply of ground water at a temperature of 18 to 25°C to stimulate hormone induced breeding of carps.

A dry methods of artificial fertilization will be provided for an equipment will include suspended nets for incubation of fertilised eggs in indoor cisterns. The cisterns will also accommodate the hatchlings for the period until they are placed in the ponds.

If, these requirements call for hatchery ponds with a general arrangement substantially different from those traditionally constructed in India and technical experience in dealing with the requirements is therefore not generally available requires strengthening.

Components of a Carp Hatchery

The Site

Sites which enable easy fulfillment of the following objectives, naturally or in-expensively, qualify for locating a carp hatchery.

1. Ponds excavated at the site should provide a water retentive soil base, exposed by digging or transfer of top soil of the site to pond bottom and embankments.
2. The soil should possess basic nutrients and respond readily to organic and inorganic fertilization.

3. There should be a dependable source of perennially available water in adequate quantity for the size of the proposed hatchery.
4. Self-draining ponds should be used on sloping sites.
5. The physical and chemical properties of the water are within acceptable limits, such that water quality can be further manipulated by chemical treatment to suit aquacultural needs.
6. The site is easily accessible by rail and/or road and air.
7. There is a market in the vicinity.
8. Fertilizers and raw material for feeds required for aquaculture operations and building material for constructing the hatchery are available near the site.
9. There is no industrial, domestic or pesticide pollution at the site.
10. There are reasonable educational and medical facilities available in the vicinity of the site.
11. There may be scope for integration of aquaculture with agriculture, horticulture or floriculture at the site.

Soil Quailty

The biological productivity of a natural pond mainly depends on the quality of its soil base. A satisfactory pond bottom soil is one which, apart from being impervious to water, permits rapid mineralization of organic matter, absorbs nutrients loosely bound and releases them slowly over a long period. Silty clays, clay-loams, loams, etc., generally make good quality soils for a fish pond. Rocky outcrops, shale ledges, sand, gravel and limestone areas must be avoided. If, however, a measure of soil porosity becomes unavoidable, then pond bottom may be treated with bentonite, clay or other soil sealants. Sprayed-on asphalt liners and plastic film liners can also be used to reduce or prevent seepage but any such treatment apart from being expensive, in effect, seals off the soil-water interface with an inert substance and prevents soil-water exchange of minerals and nutrients, which is detrimental to biological productivity. Fertilization, especially organic manuring over a period of time, automatically reduces the rate of seepage by sealing soil pores. It also reduces water turbidity caused by suspended silt and colloids. Algae, in the presence of electrolytes, aid in the latter process by flocculation. If a liner is to be installed to prevent seepage, it is desirable to put it about 200 cm below the pond bottom so that there exists a thick substratum of water-soaked soil at the bottom of the pond.

The minerals and nutrients required for securing biological productivity for the pond have basically to be drawn into the pond water from the pond soil. The pond has to be enriched artificially for sustenance of its productivity once its inherent fertility is used up. Economic considerations in aquaculture demand that the barest minimum of fertilization be done artificially. This brings to the fore the extreme importance of pond soil chemistry and its intimate direct relationship with pond fertilization.

Water Quality

Water of desirable quality and quantity is perhaps the most important requirement of a carp hatchery. The usual sources of water for a carp hatchery are rain water, reservoirs, rivers and streams, springs, irrigation canals, surface run-off, open wells, tubewells and artesian wells.

Flow-wise, by virtue of their elevated locations, perhaps the best and most reliable sources of water supply are reservoirs. Water from rivers, canals and surface run-off sources, apart from being prone to flooding (floods are often destructive), more often than not carry a heavy load of silt which is very undesirable, especially in the hatchery proper. In the hatchery proper, silt smothers eggs, hampering their development; in ponds, it leads to siltation reducing pond volume, obstructs penetration of sunlight and adversely affects pond productivity.

Water from underground sources is generally free from biota and in that respect is the safest, but often suffers from the serious drawback of deficiency in dissolved oxygen. Water with a high dissolved oxygen content (6 – 9 ppm) at required temperature is the most essential requirement of a hatchery proper. Dissolved oxygen level in the water is made good by installing aeration devices before its entry into the hatchery proper.

General Planning in Carp Hatchery Construction

Before hatchery construction commences, it is essential that the site be examined carefully as to the characteristics on various points given in the Table 9.1 below and those stated above. It is necessary to dig 2.5 m deep pits at fairly close range along a grid and examine soil samples drawn from 25 cm depth level profiles on their physical and chemical properties. Laboratory seepage tests with soil samples may be carried out to assess water retention. It is also necessary to look for depth of the subsoil water table during the hottest part of the year and carry out laboratory examination of the physical and chemical properties of the water available at the site. It is absolutely essential that water sources at the site be carefully examined as to the quantities of water available in different months of the year commensurate with the needs of the size of the hatchery desired. A detailed contour survey of the site is an essential prerequisite for preparing a master plan of the layout of the hatchery.

Table 9.1: Characteristics of a Satisfactory Carp Hatchery Site

Characteristics	*Quality*
Nature of terrain	Non-rocky having atleast 2.5 m deep soil cover.
Slope of the terrain	Gently sloping such that terraced self-draining ponds can be carved out.
Physical quality of soil	Soil fraction should be 90 per cent of the whole-soil-stone and gravel not exceeding 10 per cent.
Chemical quality of soil	pH near neutral; total available N>0.1 per cent; total available P>0.1 per cent; total available C>1.0 per cent; should be responsive to organic and inorganic fertilization.
Rate of seepage	<1 meter/annum.
Subsoil water table	Should not be far below deepest pond bottom in driest summer month in case soil is not completely water retentive.
Water supply	There should be a dependable source of perennial water supply sufficient to meet the total water requirements of the hatchery.
Chemical quality of water	pH near neutral; temperature range 20°C to 30°C; should be responsive to organic and inorganic fertilization.

Essential Components of a Hatchery

The various essential components of a fish hatchery are :

1. Broodfish ponds to hold adult fish for spawning and serve as donors of pituitary glands and to accommodate spent females and males.
2. A hatchery proper comprising a complex of facilities for fish spawning, hatching and care of hatchlings to raise them up to postlarval stage.
3. Nursery ponds for rearing postlarvae to fry stage.
4. Rearing ponds for growing fry to fingerlings.
5. Ponds for production of fish to supply broodfish ponds and donors of pituitary glands.

In this portion, consideration is first given to the land areas to be converted in to different pond types and hatchery facilities. For this, one has to start with certain basic decisions, including the carp species, their sizes and the quantities desired to be produced. It is presumed here as an example that the species to be produced are among the species of carps mentioned earlier in this book (three Indian major carps (IMC) and three exotic carps) and that the objective is to produce annually 100 tonnes of table-sized fish of these species.

The assumptions shown in Table 9.2 are made to derive the requisite numbers of various life-history stages of different species to produce 100 tonnes of fish per year.

The following ratios of the carp species are further presumed.

Surface Feeder		*Percentage*
Phytoplankton	Silver Carp	20
Zooplankton	Catla	20
Bottom Feeder		
Omnivore	Common Carp	20
Detritus	Mrigal	10
Column Feeder		
Browser	Rohu	15
Macrovegetation Feeder		
Aquatic foliage	Grass Carp	15

Table 9.2: Assumptions of Derive Pond Areas and Numbers for a Production of 100 Tonnes of Fish/Year

Number	*Assumption*
1.	It is a system of polyculture of six species, *viz.* Catla, Rohu, Mrigal, Silver Crap, Grass Carp and Common Crap.
2.	The density of production of table-size fish is 5 t/ha.
3.	The rate of stocking of postlarvae is 1.5 million/ha or 150/m^2.
4.	The rate of survival of postlarvae to the fry stage is 50 per cent.
5.	The rate of stocking of fry to raise them to fingerlings is 125,000/ha.
6.	The rate of survival in rearing fry to the fingerling stage is 80 per cent.
7.	The rate of survival in rearing fingerlings to table-sized fish is 90%.
8.	A fish is considered marketable when it is 750 g in weight.
9.	An average female broodfish produces 50,000 viable postlarvae.
10.	Proportion of female broodfish which will respond to hypophysation is 50 per cent males 100 per cent.
11.	Average weight of female spawner preferred for hypophysation is 4.5 kg.
12.	On the average, a female broodfish for induced breeding needs 10 mg/kg of pituitary hormone and a male broodfish requires 4mg/kg of pituitary hormone.
13.	On as average a donor fish would yield 3 mg/kg of pituitary gland.

It is taken for granted that the stock ponds under consideration are still-water ponds provided with facilities for draining and, therefore, inlets and monked outlets. For production of 100 tonnes of tabled-sized fish per year, the water surface of stock ponds required is 20 ha.

The number of fingerlings (y) required per hectare can be calculated from the following formula:

$$a \times y \times b = c$$

where,

a: Survival rate.

b: Size at harvesting.

c: Yield at harvesting.

Substituting the assumed values for a, b and c one can solve for y as follows;

$$0.9 \times y \times 0.75\text{kg} = 5{,}000/\text{ha}$$

$$y = 7{,}407 \text{ fingerlings/ha}$$

The required number of fingerlings of each species can be calculated as shown in Table 9.3 below.

Table 9.3: Number of Fingerlings of Different Species to Produce Annual Harvest of 100 tonnes

Species	*Species Composition (per cent)*	*Requirement of fingerlings (no./ha)*	*Total Requirement of fingerlings (no./year)*
Silver carp	20	1,481	29,620
Catla	20	1,482	29,640
Common carp	20	1,481	29,620
Mrigal	10	741	14,820
Rohu	15	1,111	22,220
Grass Crap	15	1,111	22,220
Total	100	7,407	148,140

Fingerlings/ha × 20 ha, being the requirement of water surface area to produce 100 t/year.

Based on a survival rate of 50 per cent in nursing postlarvae to fry, the number of post-larvae required can be calculated as;

$$\frac{185{,}175}{0.50} = 370{,}350$$

Based on a survival rate of 80 per cent in rearing from fry to fingerling stages, the total number of fry required can be calculated as :

$$\frac{148{,}140}{0.80} = 185{,}175$$

As per assumption 5, to stock 185,175 fry, the rearing space required is 185,175/125.000 = 1.48 ha, say 1.5 ha. If the rearing pond is 0.2 ha, which is s a convenient size for netting purposes, eight rearing ponds will need to be prepared in the hatchery.

As per assumption 3, the stocking rate of postlarvae is 1.5 million/ha. Therefore, to stock 370,350 postlarvae, the nursery pond is 0.05 ha, which is a convenient size for netting operations, five nursery ponds will need to be prepared.

Requirements of Broodfish Ponds

All the carps under consideration here have high fecundities which vary a great deal according to species. Further, fecundity is a function of the age of the broodfish, its nutrition and, therefore, the feeds it has been given. However, for the sake of breeding purposes, and until enough scientific knowledge of the control of fecundity in different species of carps is gathered, female broodfish weighing 3-6 kg each can be presumed to produce 50,000 postlarvae each (Assumption 9). Furthermore, a set of spawners of a given species is taken to comprise three specimens one female and two males, such that the total weight of female broodfish approximately equals the total weight of male broodfish.

In the present illustration, the number of postlarvae required is 370,350. This number of postlarvae can be produced by induced breeding eight female spawners of different species together weighing 36.0 kg (4.5 × 8 = 36.0) and 16 males also together weighing 36.0 kg. But since there are six different species, two female broodfish of each species would be required, which, at an average weight of 4.5 kg, would altogether weigh 9.0 kg/species. For six species, the total weight of females would be 63.0 kg. But vide Assumption 10, success in production of postlarvae may be only 50 per cent. Therefore, the total weight of female broodfish required in this model is 126.0 kg.

The approximate number of male broodfish required remains 28, together weighting 63.0 kg. The total weight of female and male broodfish would thus be 189.0 kg.

Based on Assumption 12 of hormone requirements, 126 kg of female broodfish would need 1,260 mg (126 × 10 = 1,260 mg) of pituitary gland and 63 kg of male broodfish would need 252 mg. Total weight of pituitary glands required would be 1,512 mg.

Based of assumption 13 of 3 mg/kg of pituitary gland from donor fish, 1,512 mg of pituitary gland material can be got from 554 kg of donor fish taking into account 10 per cent wastage.

Requirements in weight of adult fish for breeding as well as for donor fish, therefore, are 743 kg (189.0 kg for breeding and 554 kg for donor fish).

The rate of stocking of broodfish ponds is 1,000 kg/ha. Therefore, for stocking 743 kg of adult fish 0.743 ha or say 0.8 ha are required. It is improper to have a broodfish pond of less than 0.2 ha each for convenience of netting, feeding, aeration, application of prophylactic and therapeutic chemicals as well as flushing the pound with fresh cool water for proper gonadal development. Therefore, in this illustration, four broodfish ponds of 0.2 ha each are to be prepared, which will all together given a stocking capacity of 600 kg of adult fish against the requirement of 449 kg. It must, however, be borne in mind that when the targets of production are high, many broodfish ponds, which should be different from stock ponds, would be required. For further information of current practices in broodstock husbandry (stocking density and nutrition) are given below :

Broodstock Husbandry	Nutrition
All carps : kept in manured ponds with supplement feeding; various polyculture combinations; total stocking density 1,000-2,000 kg/ ha.	Supplemental feeding at about 1 per cent body weight/day with rice bran : oilcake (1 : 1).

Much of this information is incomplete since culturist frequently change practices based on experience to cope with different species combinations and local conditions.

Twenty-eight female spawners, allowing for 50 per cent success in ovulation by hypophysation, complemented by 28 male spawners, would besides supplying 1,512 mg of pituitary glands for hypophysation, on induced breeding, furnish 370,350 postlarvae. These would need 0.25 ha of nursery space and would, at the survival rate of 50 per cent, yield 185,175 fry. These fry, at the survival rate of 90 per cent would produce 148,140 fingerlings requiring 1.6 ha of rearing ponds. The total of 743.0 kg of broodfish of all the seven species required to produce 370,350 postlarvae, would require 0.80 ha of broodfish ponds.

Table 9.4: Requirements of Different Kinds of Ponds for Production of 100 t/year of Table-Size Fish

Types of Ponds	*Area of Ponds (ha)*	*Number of Ponds*	*Total Area of Ponds (ha)*
Broodfish	0.20	4	0.80
Nursery	0.05	5	0.25
Rearing	0.20	8	1.60
Stock	1.00	20	20.00
			22.65

Structural Features for Different Types of Ponds

The salient feature of different types of ponds are :

Nursery Ponds

Drainable, small ponds 0.01 to 0.1 ha and depth around 0.5 to 1.2 m. pond bottom gently sloping towards outlet but plane. Rectangular shape and convenient width for netting operations. A sump at outlet for collection of fry.

Rearing Ponds

Drainable ponds larger than nurseries and of depth of around 0.8 to 2 m. Pond bottom plane but gently sloping towards outlet. Rectangular shape and convenient width for netting operating. A sump at outlet for collection of fry and fingerlings.

Drain ability of both nursery and rearing ponds highly desirable because during the slack season, it would enable the pond bottom to be exposed to sun, aeration mineralization and aerobic destruction of the organic load which otherwise might lead to hydrogen sulphide production, which is toxic. Other advantages of drainability area ease of control of parasites, predators and unwanted fish; easy of application of toxicants, manures and fertilizers of ploughing is necessary, facilitating cultivation of nitrogen fixing crops like leguminous plants; easy of harvesting, a process greatly facilitated by constructing a sump at the approach to the outlet.

Stock and/or Broodfish Ponds

Drainable, of rectangular shape should have provision for flushing with 'cold' water of 25-28°C. A convenient source of such water is a well (artesian, open or tube) or a underground spring. A width of 20-30 m facilitates netting operations. Depth are generally 1 to 2.5 m but greater depth is preferred in cold climatic zones and arid or semi-arid areas where the rate of evaporation is high.

In this case of nursery and rearing and even stocking ponds, it is advisable for each pond to have its water supply routed through a sand and pebble filter set up in a concrete structure with water moving upward from below (reverse flow filter) and entering the pond from the top of the filter. The advantage of individual water filters for each nursery and rearing pond is that the filter would not only clean the water but very effectively control the entry of unwanted fish, their young and even the eggs, into the ponds. One of the more important causes of mortality of carp hatchlings and fry in nursery and rearing ponds is predation by unwanted fish and the filter would provide the most effective way of removing the cause of the problem. The same applies to stock ponds where carp fingerling can be preyed upon unwanted fish. The unwanted fish not only prey upon carps but also share the natural and artificial food which is meant for carps.

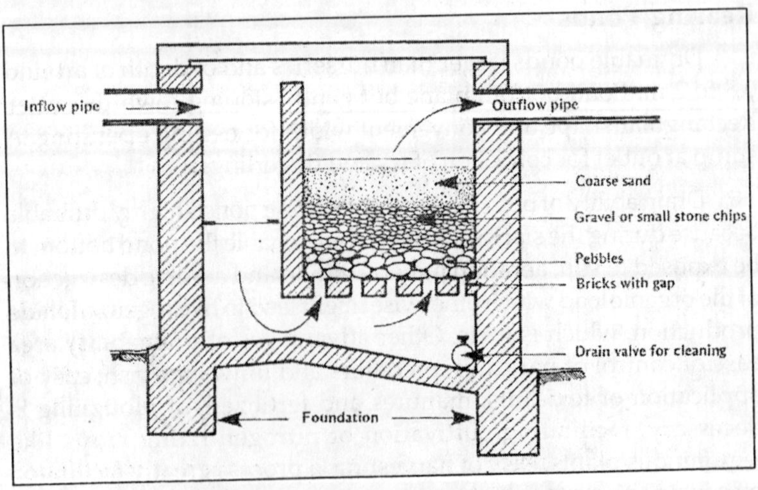

Figure 9.1: A Reverse Flow Filter (Reproduced from Woynarovich 1975)

Placement of Different Types of Ponds in a Hatchery

This location of different kinds of ponds in a hatchery is of considerable importance for ease of operation and minimization of operating costs.

As mentioned earlier, a detailed contour survey of the area where the hatchery is to be built must precede fish farm construction. This will minimize filling or digging and movement of earth, which all cause expenditure on labour or mechanical earth moving. The lowest area of the terrain must be developed into stock ponds, appropriate higher areas developed into rearing ponds and areas higher still into nursery ponds. The same principle would apply to developing swampy and marshy lands into fish farms, the deepest being converted into stock ponds. The highest of the areas, on this principle, are to be developed into the hatchery proper, closest to which must be located water treatment filtration and sedimentation plants and the water tower, laboratory, hatchery, office and residential houses for the staff. This arrangement will place the nurseries closest to the hatchery proper, which is a logical arrangement. The fish breeding tanks from inside the hatchery building may lead to outdoor nursery ponds by suitable conditions. As hatchlings grow into fry and fingerlings, they can be with least expense, conveniently transferred from nursery to rearing to stock ponds. Broodfish in the ponds in this above state arrangement would be farthest from the hatchery, however, the ante-tanks in the hatchery provide holding space for broodfish close by the fish breeding tanks.

Figure 9.2: A Bird's Eye View of the Fish Hatchery and Training Centre Raipur, Noekhali, Bangladesh

The Hatchery Proper

The hatchery proper is perhaps the most vital component of a modern fish farm. Incorporation of a hatchery in a carp farm is a relatively recent development in the history of carp culture. It is logical corollary of the development of induced breeding as a technique for commercial production of carp seed. In providing flowing water of high oxygen content, a carp hatchery follows the basic principles of a salmonid hatchery in which stripping has been a very old practice in different parts of the world. In the early days of induced carp breeding, there were only outdoor breeding and hatching facilities mainly comprising cloth enclosures (called *hapas*) installed in ponds. The use of *hapas* involves mechanical transfer of eggs from breeding *hapas* to hatching *hapas* and from hatching *hapas* to storage *hapas* for temporary storage while they grow to an early fry stage before transfer to carp nurseries for raising to the late fry stage.

The essential components of a modern hatchery proper are ante-tanks or storage tanks; breeding tanks or ward tanks; incubators of hatching jars; larval rearing tanks. The infrastructural facilities and functional alternatives of these components, as they have evolved with the passage of time, are described below.

Figure 9.3: A Breeding Hapa

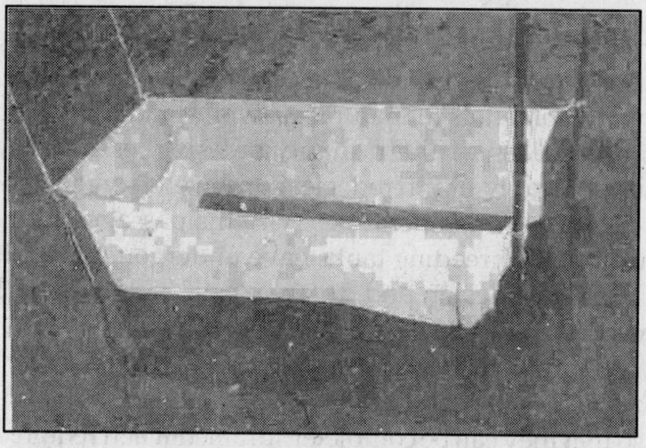

Figure 9.4: Hatching Hapa

Ante-Tanks

The purpose of ante-tanks is to hold selected ripe breeders prior to hypophysation. The ante-tanks must be at least two in number, one for each sex of species of fish in the smallest hatchery. A 200 m^2 or 450m^2 (10m × 20m or 15m × 30m) ante-tank of 1.0 to 1.5 m depth can hold 25-50 sets of broodfish (each set comprising one female and two males) with female spawners weighing 3-6 kg each and sexes segregated in the two tanks.

In many modern hatcheries, ante-tanks may serve diverse functions at different times. These functions are :

1. holding breeding before hypophysations;
2. holding and rearing fry prior to sale;
3. holding fingerlings prior to sale;
4. serving as treatment tanks for diseased or infected fry, fingerlings and broodfish;
5. providing additional space for operating jar or funnel incubators.

In case the ante-tanks are to serve as multipurpose tanks in a hatchery, their dimensions may be altered to a more elongated shape, retaining their twin character, and their number increased as per need.

Breeding Tanks

Also called ward tanks, the purpose of breeding tanks is to hold injected breeders for natural spawning and fertilization. These tanks, regardless of their shape and size, essentially need a continuous supply of filtered, clear and well-oxygenated running water of optimum temperature. With the passage of time in carp aquaculture, the breeding tanks have undergone great changes in their shape and design and, to an extent, 'automation' has been incorporated to effect self-transfer of fertilized eggs into hatching tanks. A common feature of breeding tanks regardless of their shape and size, is that they should have a sloping bottom leading to the outlet so that they can be completely drained when required without leaving any eggs behind. Another common feature of breeding tanks is that it should be possible to maintain their water level. This is universally done by installing an outside standpipe (also at times called turn-down pipe) in each hatching tank.

A breeding tanks may be rectangular in shape, a convenient size being 2.5m × 1.5m × 1.0m for holding 4-6 breeders and 4m × 2m × 1m for 8 – 10 breeds weighing 3 – 6 kg each (both female and male) in each case. For still larger breeder weighing 12-20 kg each, even bigger breeding tanks, 7.5m × 2.5 m × 1.0 m may be made. In breeding tanks of rectangular shape, if stripping is to be done, it is convenient to construct a 20 – 25 cm deep and 50 cm wide side-tank along the long axis of the breeding tank. The fish to be stripped may be conveniently pushed into the ditch without much disturbance to the remaining fish in the breeding tanks. Stripping operations inevitably have to be done outside the tank in the hatchery building. It is advisable to fix a breeding *hapa* inside a breeding tank, the rectangular shape of which is more conducive to installation of a *hapa*. An advantage of *hapa* in the above context is that it is much easier to handle, procure of release a spawner from and into a *hapa* and also to insolate and pick up individual spawners should it become necessary. The latter becomes quite difficult in a community breeding tank whatever its shape, despite operating a dip net.

Circular Breeding Hatchery and Larvae Tanks (Spawning Pool)

In modern times, more and more carp farms are incorporating circular breeding tanks in their hatcheries either as substitute for existing rectangular breeding tanks as additional facilities. Breeding tanks of circular shape are of Chinese origin. The advantages of a circular breeding tanks are :

1. It is conducive to a continuous flow system.
2. Within a limited space, it provides the effect of some aspects of the riverine environment which is the natural habitat of carps.
3. The centrifugal flow make the operating of inlet and outlets more effective.
4. There are no dead areas as far as water flow is concerned, such as one encounter in rectangular or polygonal shapes; this make distribution of oxygenated water in the tanks more uniform and, hence, hatching more effective.
5. It is more conducive to protecting eggs from being washed out of the hatching facility by installation of a screen surrounding the centrally located outlet.
6. It make additional aeration easier outside the centrally located outlet.
7. It can combine the function of a breeding tank, a hatching tanks as well as larvae-rearing tanks.
8. It can lead itself to a modular type of design in which each breeding tank can be linked to twin circular hatching tanks such that the entire contents of the former can be accommodated at a time in a pair of hatching tanks with water containing fish eggs flowing by gravity. This procedure enables performance of a series of breeding operations almost continuously in the fish breeding season, given adequate number of breeding and hatching units of modular design.

The drawbacks of a circular breeding tanks are that for functional effectiveness in order to breed 4–6 sets of broodfish weighing 3–6 kg each, a circular tank, by virtue of its shape, has to be larger in volume than a rectangular tank. Hence, it needs more water and in many hatcheries, water is a serious limiting factor. Monitoring of the extent of fertilization and hatching in fish breeding operations become difficult, especially in large circular breeding tanks. Defects in leveling in the drainage system especially links with hatching tanks, becomes difficult to repair and remedy without dismantling the centrally located outlet structure underground below the brick work of the hatching tank.

Spawning Pool

In only days spawning best used to be of the sing of 8 m in diameter with a depth of 1 m from the top of the pool. This pool will be provided with 8-duck mouths at the bottom and 8-duck mouths in the middle of the wall of the spawning pool. All the duck mouths will release water simultaneously in the clockwise direction. In the bottom of the pool, slopes down to the center at 1.5 m water supply to this pool is from the overhead tanks by means of 7.5 cms diameter pipe. The water supply line is laid along the outside of the wall and inlet to the pool is provided at 14 to 16 places equally spaced and fixed at an angle of 45º to the radius of the tank using 20 mm diameter with nozzle mouth all in one direction. They are fixed to the vertical wall, the nozzle mouth is flush with cement plaster faces and near the bottom along the periphery of the pond. In the centers an outer

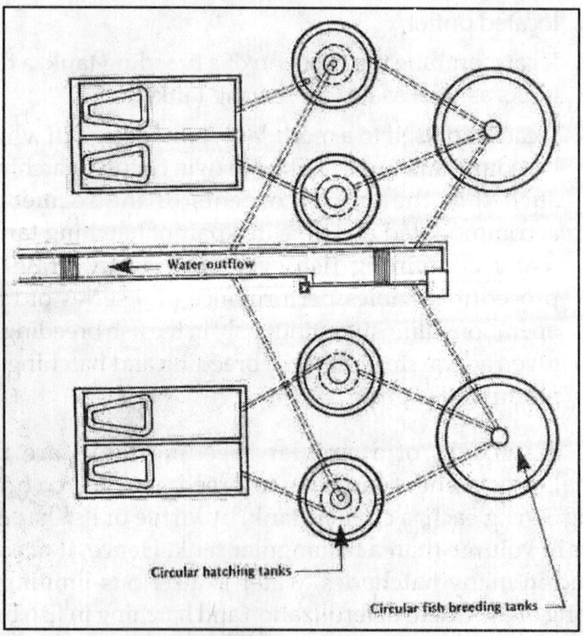

Figure 9.5: Circular Fish Breeding, Hatching and Larvae Rearing Tanks in a Modular Design

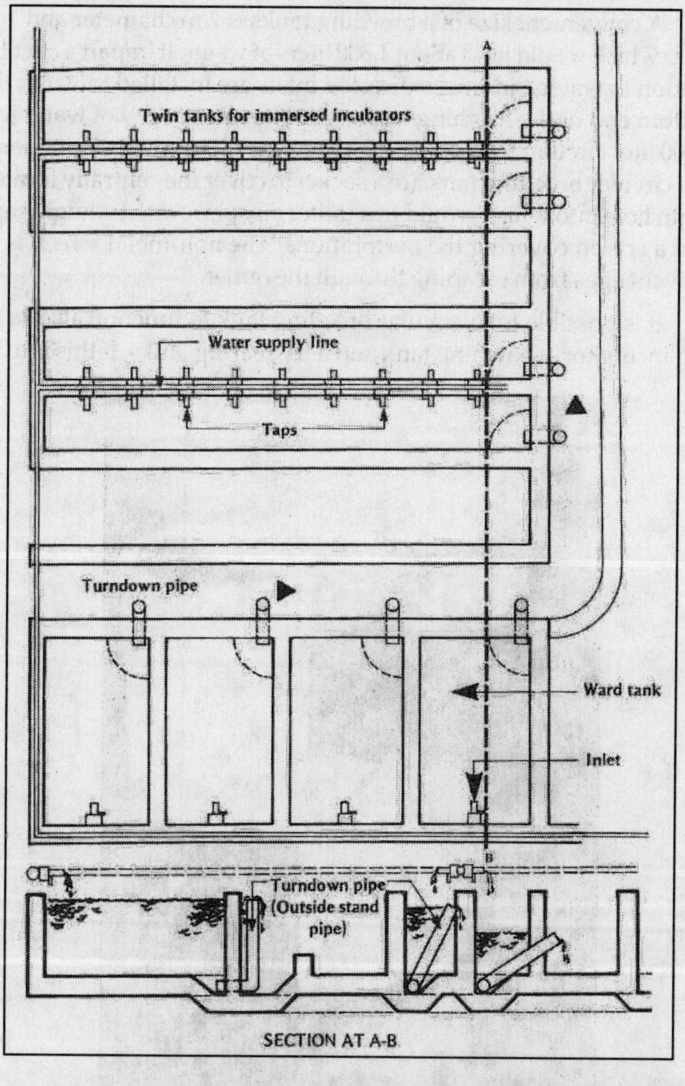

Figure 9.6: Twin Tanks and Wards Tanks in a Hatchery; Also Seen are Outside Stand Pipes or Turn-Down Pipes (Reproduced from Woynarovich and Horvath 1980)

pipe of 10 cms diameter is fitted through which, on opening the valve fertilized eggs alongwith water are transferred into hatchery or incubation pond for hatching.

A convenient size of a breeding tanks is 2 m diameter and 1 m deep which would hold about 1,800 liters of water. It impart a circular motion to water, sideways-directed inlets are installed towards the bottom end of the hatching tank well. The rate of flow of water in a 1,800 liter circular tank may be 6-8 liters/sec. The other components of a circular breeding tank are a socket to cover the centrally located drain hole into which would fit a uniformly perforated straight pipe and a screen covering the perforations. The non-metal screen is to prevent eggs from escaping through the outlet.

It is possible for a circular breeding tank to function also as an egg incubator, a hatching tank and a fry rearing tanks. If this is to be

Figure 9.7: A Circular Concretes Fish Breeding Tank; Also Seen are a Sideways Directed Inlet and Centrally Located Drain Hole Over which is Fitted a Straight Perforated Pipe Covered by a Screen (Reproduced from a photos taken at the fish Hatchery and Training Centre, Raipur, Noakhali, Bangladesh)

Figure 9.8: Carp Eggs Incubating in Zoug Jars at a Fish Hatchery in Dinnyes Hungery (Reproduced from Huest 1972)

done, then the eggs must be allowed to remain in the breeding tank after they are spawned and fertilized and additional aeration given to the them. This can be conveniently done by installing an air-diffuser at the base of the screen surrounding the straight perforated pipe. The air diffuser should be of circular shape to encircle the perforated metal and be connected to an air blower or compressor through rubber tube. The water level regulation in the circular breeding tanks is done by the vertical stand pipe tightly fitting the central socket and through the turn-down pipe, manipulable from outside the circular breeding tank.

Hatching or Incubation Pond

There are two circular incubation or hatching pools 3.6 m internal diameter. There are two chambers in each pool. The outer dimension is 4 m having outer masonary concrete wall. Another circular wall with fixed mesh screen is provided at 0.74 m clear distance from the outer wall. The inner chamber is provided with 10 cm diameter vertical outlet with holes at different heights for taking out excess water or dewatering of the hatching pool. The spawn alongwith water flows from these pools to spawnery (spawn collection pool).

Water Connection from Overhead Tank

From the overhead RCC tank 7.5 cm diameter pipe line reduced to 5 cm diameter pipeline and then to 1.2 cm diameter pipelines, 8 numbers of outlet are fitted in the floor having duck mouth opening fixed in one direction at an angle of 45° towards the inner wall.

Water supply pipe will be fitted from the circular spawning tank by 10 cm diameter pipe line and then bifurcated into 2 number. 5 cm diameter pipe lines, one in each of the hatching pool which are further connected to duck mouth in the floor of incubation pond.

There should be an outlet of 7.5 cm diameter through which hatchlings pass to receiving pond. This opening may also be used for complete dewatering of outer chamber of the hatching pool. Desired water movement is about 0.2 to 0.3 m/sec.

Spawnery

This is a rectangular tank. The inside dimensions are 4 m × 2.5 m × 1.2 m depth. This is located at a lower elevation than the hatching pool, so as to drain out the water from it by gravity, if ground level may permit.

Water Supply

Water supply from overhead tank by 7.5 cm diameter pipeline bifurcated into 3 numbers. 5 cm diameter pipe is given in the form of spray for aeration. From each of the hatching pool 7.5 cm diameter are provided for transfer and regulation of spawn intake into spawnery.

The overflow from the spawnery is to be discharged into the open drain and suitably utilized in the pond near by.

Overhead Tank

The floor of the tank is to be 2.6 meter above floor level at center of spawning pond. The inside dimension can be kept 5.5 m × 2.7 m × 2.2 m. Height and having a capacity of 30,000 liters. Water supply to over head tank is to be arranged by pumping water from a deep tubewell.

Pond

In a Hatchery, following ponds need to be constructed.

1. Brooder Pond
2. Nursery Pond
3. Fry Stunting Pond

Brooder Pond (Size 100 m and 50 m)

As the purpose of the fish farm would be to produce fish seed and also the table size fish, the average size of the pond will be maintained to be of 0.05 hectare measuring 100 m × 50 m (rectangular in size).

This type of pond will serve the purpose of fry raising to table-size fish well as maintenance of "Brood-Fish" for breeding purpose.

Nursery Pond (Size 60 m × 40 m)

The Spawn and fry produced in the Hatchery would be kept in this pond atleast for 14 days and only after the completion of 14 days, the fry will be transported or sale either on the spot or to the consumer's pond.

Fry Stunting Pond (Size 30 mtrs. × 20 mtrs)

This type of pond will be prepared by brick soling on the bed as well as on the lateral side of the embankment. The unsold spawn or fry will be kept in this pond with little amount of feeding. This fry will remain stunted and well not grow very fast. The stocking material will be available throughout the year from this pond. The fry may be kept up to 10 months in the stunted conditions throughout the year.

As soon as, the market demand will require, even during the off season, when the fry will not be available anywhere, the stunted fry will be supplied to the farmers throughout the year also such stunted fry will be utilized for stocking the farm pond as and when required.

As the fries are being kept in a stunted condition in a limited space with a meager supply of food, the growth will not be faster and the fry will remain in stunted condition. As soon as, such fries are released in the larger pond, the growth would be very fast and even a stunted fry can attain the size one kg in six months.

In this way, two crops can easily be obtained from a single pond with an average production of 6-7 tonnes per hectare.

Operation of the Circular Hatchery

1. Brooders are put in the spawning pool for about 4 to 8 hours for conditioning. Between 4 to 6 p.m. first injection is given to the females. At a time 40 kgs of females and 60 kgs of males are put in a spawning pool. Then after six hours, second dose of injection is given to the female and one dose 3 – 4 mg/kg body weight to the male. After about one hour of the second injection, the water jets are started so as to get the circular motion in the water. After 6 to 8 hours of second injection, breeding takes place. Over crores of eggs can be collected in one operation only.
2. The eggs are collected at the bottom. As the bottom of the spawning pool slopes to the centre, where there is an outlet pipe leading to the collection chamber. The collected eggs are transferred into the hatching pool through pipe by opening the valves.
3. Arrangement will be made to churn the water again in the hatching pool in four days time, the spawn will be about 6 mm in size and then it will be taken into spawnery.

From there it is lifted and stocked in separate earthen ponds for growth or, if there are surplus spawn produced it would be kept in the "fry stunted pond" already described above.

If oxygenation is less, aeration can be given through compressor or aerator in the hatching pool. For aeration water showers, water jets will also be provided.

Economics of the Operation

During the breeding season lasting for about 120 days in a year, the breeding and hatching operation can be carried out into 30 batches. Each batch of 4 days. About one crore (10 millions) eggs can be hatched in one batch (spawning pond 8 m dia and hatching pond 3.6 m dia) and with 95 per cent hatching success to 85 million spawn can be produced. The size of which would vary from 6 to 8 mm. With the sale price of Rs. 100/- per thousand and the income comes to Rs. 2,85,00,000/-. Deducting about 50 per cent cost per annum on an average for capital investment, maintenance, supply of brood stock and miscellaneous expenses net profit of Rs. 1,42,50,000.00 would be achieved during one season subject to the condition that proper arrangements are made to carry out uninterrupted operation of the hatchery during the breeding season. But assuming that about 15 batches could easily be handled on an average due to non-availability of matured brood stock the net profit would be about Rs. 71,25,000.00 per breeding season per hatchery.

Economics of Hatchery
(Approximate cost estimate of circular hatchery (2010–2011 base))

Sl.No	*Item*	*Quantity*	*Cost (in Rs in Lakhs)*
1	Spawning pool – brick cum RCC pool of 8 m dia and 1.5 m average depth	1 Nos	9.50
2	Hatching pool – brick cum RCC pool of 4 m dia, having another size chamber with screen depth 1.2 m.	2 Nos	6.00
3	Spawnery – brick cum RCC wall 4 m × 2.5 m × 1.2 m height	1 Nos	1.00
4	Overhead water storage tank bottom 2.6 m above ground level and capacity of 30000 litres	1 Nos	4.00
5	Tube-well with water supply lines	1 Nos	0.50
6	Screen for spawning pool with removable pipes	1 Nos	0.050
7	Operational cost		0.050
8	Miscellaneous expenditures such as fencing, electric service line, side clearance area 25 m × 25 m, drainage, approach road etc.		1.00
9	Contingency @ 5 per cent		1.105
10	Establishment @ 2 per cent		0.442
	TOTAL		23.647

Description of Area

1.	Land Area	12 Acres = 522720 sq.ft
2.	Location	Near village Porran Tar, West of School North of Canal, West and North Agricultural land.
3.	Water Source	As per report of the villagers, the boring well indicates the water table to be 80 ft. below ground level.

Contd...

4.	Soil Condition	From the physical examination of the soil indicates that the soil is of the silt base. Five pit holes were dug out at various places randomly selected in around the available land which clearly shows that the soil profile is homogenous (silty) upto 12 m from ground level. After 1.2m finer sand particle constitute the major part.

On the date of 8th January 2010, the water started oozing out only after a depth of 1.5 m which clearly indicates that there would be no shortage of water to the proposed site.

Soil

The soil samples were collected from different depth, surface soil 0.60 m below surface and 1.2 m below surface and sent to the Bihar College of Engineering, Patna for analysis of their grain size and permeability co-efficient. On the basis of the report to be obtained from the Engineering College, the final feasibility report regarding the suitability of the site for the construction of pond for raising 'Fish Seed' and 'Table Sized Fish' will be finalised.

Incubators or Hatching Jars

Unless the functions of a hatching tank are combined with that of a breeding tank, separate hatching facilities are required in a hatchery. A wide range of options facilities are required in a hatchery. A wide range of options of a available to fish farmer in incubating and hatching facilities.

Carp eggs can be incubated in carp hatchery jar, the water inlet can be from below or above but the exit is always at the top. The design of the jar would depend upon which type of inlet is chosen. Ordinarily, however, hatchery jars with water supply from below are more commonly used. The hatchery jars with water supply from below can be of diverse volumes and shapes. The shape may vary from plain cylindrical contour to a funnel conical or barely shape. The material can be clear glass or clear plastic. The volume of individual hatching jars may vary from 1 liter to 200 liters. A one-liter jar has the capacity to hold 100,000 water-hardened swollen carp eggs.

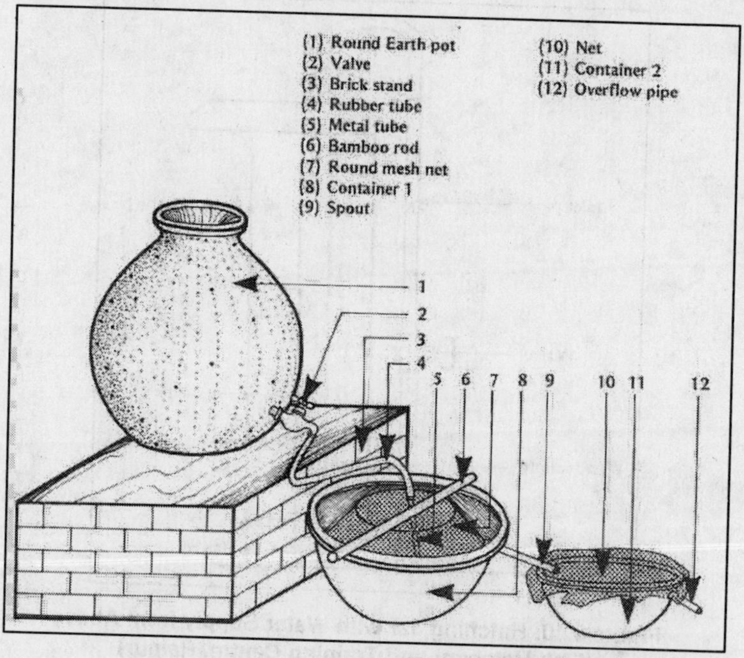

Figure 9.9: Baked Clay Vessels Serving as a Carp Hatchery in a Rural Area in India (Reproduced from Central Institute of Fisheries Education (1983)

Funnel-Shaped Hatchery Devices

In yet another hatching arrangement, soft material lie nylon or canvas is used to prepare funnel-shaped, immersed type of incubators which are installed in rectangular twin basins with water entering each incubator from the bottom. The water supply line, with tape fixed on both sides, is placed on top of the dividing wall. The distance between taps may be about 60 cm. The supply line may be of 5 cm diameter for 20 taps and 7.5 cm wide for more than 20 taps. The twin basins may be 6-8 m long, 70 cm wide and about 1 m deep and have a turn-down pipe at one end to regulate the water level.

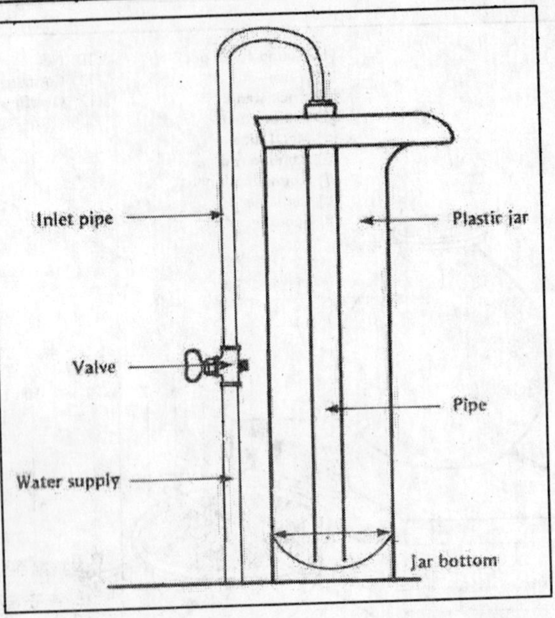

Figure 9.10: Hatching Jar with Water Supply from Above (Fish Hatchery and Training Centre, Raipur)

Figure 9.11: An Overall View of a Glass Jar Hatchery at Cuttack, India (Reproduced from Bhowmick 1978).

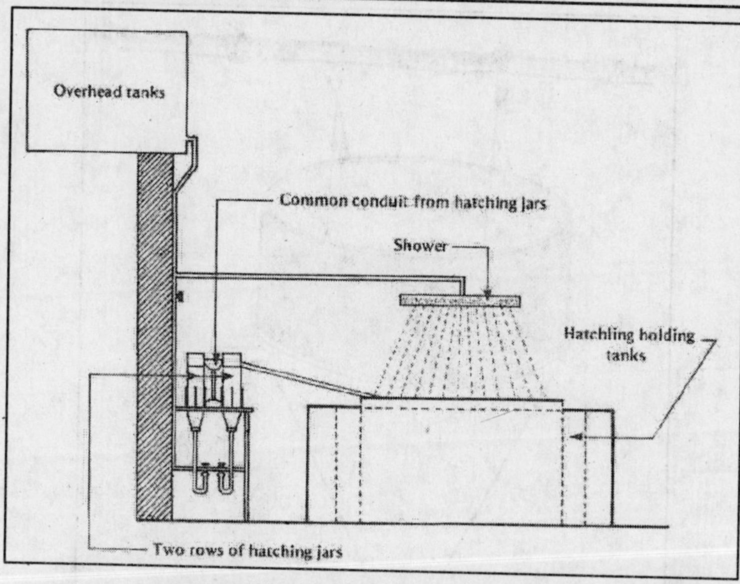

Figure 9.12: Placement of Hatching Jars in Rows of Two atop a Rectangular Table with a Common Conduit from Jars from Each Row (Reproduced from Bhowmick 1978)

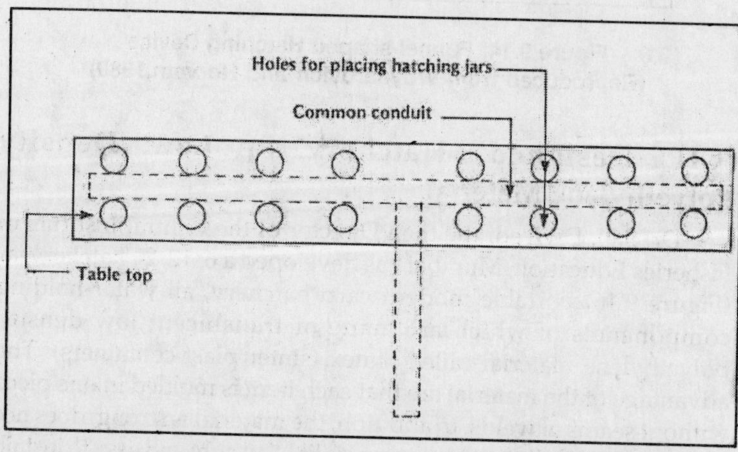

Figure 9.13: Glass Jar Hatchery Showing Drainage from Hatching Jars through a Common Conduit into a Hatchling Holding Tank with Shower Arrangement (Reproduced from Bhowmick 1978)

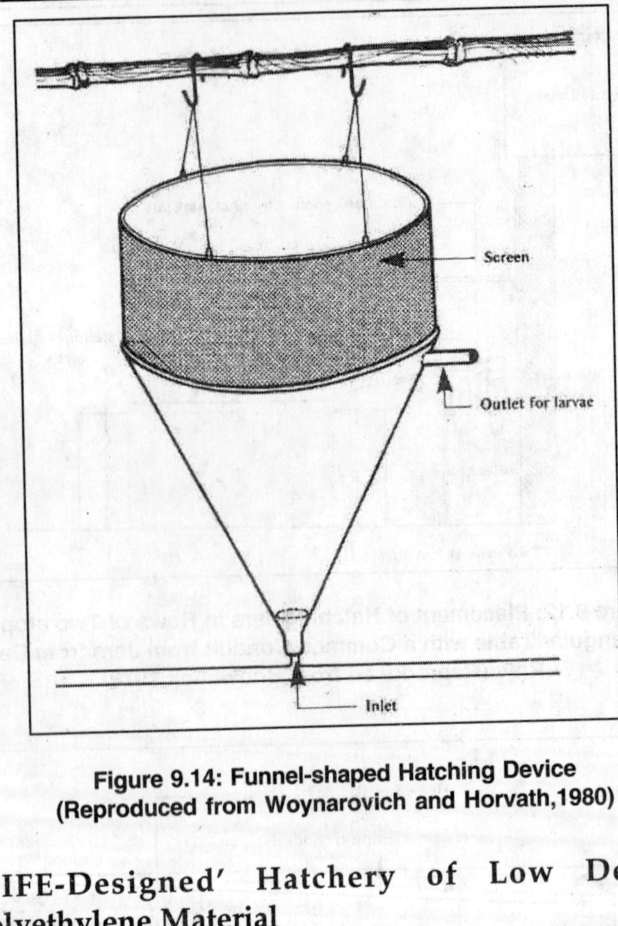

Figure 9.14: Funnel-shaped Hatching Device (Reproduced from Woynarovich and Horvath,1980)

'CIFE-Designed' Hatchery of Low Density Polyethylene Material

Dr. S.N. Dwivedi the then Director of the Central Institute of Fisheries Education, Mumbai has developed a 6, 12, 24 hatchery jar (Figure 9.16) portable modern carp hatchery, all water-holding componenets of which are made of translucent low density polyethylene material called Sintex (sinter plast containers). The advantage of the material are that each item is molded in one piece without seams of welds. In addition, the material is strong, does not corrode, requires little maintenance, is light-weight and easy to install, clean and move. This carp hatchery consists essentially of a breeding and a hatching unit. The Breeding units comprises a cooling tower, large pools with spray shower and a water circulatory system, the

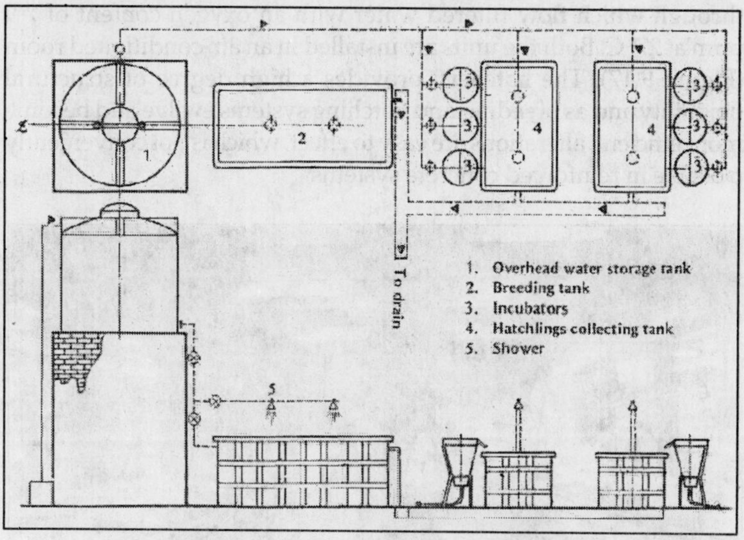

Figure 9.15: Schematic Diagram of a Low Density Polyethylene Made Six-Incubator Hatchery Unit

system furnishes clear, cool, highly oxygenated water with a gentle current from which metabolites are continually removed. The hatchery unit comprises vertical hatching jars each of 40 liter capacity

Figure 9.16: Hatchery Equipment Made of Low Density Polyethylene Material Showing Portable Hatchery Jars and Hatchling Containers among Others

through which flow filtered water with an oxygen content of 7-9 ppm at 27º C. Both the units are installed in an air-conditioned room (Figure 9.17). The hatchery provides a high degree of structural flexibility and as breeding and hatching systems evolve and become more efficient, alterations are easy to effect, which is not conveniently possible in reinforced concrete systems.

Figure 9.17: General View of a Modern "Dwivedi" Carp Hatchery Operating at Khutela Bhata Fish Seed Farm, Madhya Pradesh, India (Reproduced from Central Institute of Fisheries Education 1979-1980)

For operating hatching jars having water supply from above, jar tank units (Figure 9.18) may be intalled. A jar tank unit consists of a rectangular brickwork cemented tank (4m × 1m × 3/4m) to hold about 3,000 liters of water. Over the brickwork cemented tanks may be placed planks in pairs to support a certain number of hatchery jars also arranged in pairs, with a discharge arrangement into the brickwork cemented tank located directly below the wooden planks (Figure 9.19).

The various hatching methods described above have their own merits, demerits and limitations. These are mentioned in Table 9.5 in a comparative manner.

Table 9.5: Merits and Demerits of Different Hatching Methods

Hatching Method	*Merits and Limitations*	*Demerits*
Double walled hatching cloth hapa installed in still-water ponds	1. Useful for hatching relatively small number of eggs. 2. Separates egg shells from hatchlings very effectively.	1. Water quality connot be controlled, exposed to outside environmental conditions which may lead to mortality. 2. Fish from water outside the hapa can suck of otherwise damage and even destroy eggs and hatchlings-through the hapa wall. 3. Needs frequent replacement because cloth submerged in water gets spoiled.
Funnel-shaped canvas hatching devices provided with flowing water	1. Small scale hatching operations are possible. 2. Water quality can be controlled.	1. There is a limitation as to volume of individual funnel 2. Developing eggs are not subject to view and easy examination. 3. Discarded egg shells and hatchlings remain mixed up in the some space.
Jars with flowing water supply from below	1. Can be of diverse volumes and hence adaptable to large-scale operation. 2. Eggs and larvae are subject to view and easy examination. 3. Do not require much water. 4. Of simple design and less expensive. 5. Water quality can be controlled.	1. There is a limit to magnitude of operation 2. Need constant vigil and care. 3. If made of transparent plastic, may, with use, become translucent and opaque and then would need replacement. 4. Can break if made of glass unless of non-breakable glass which is very expensive.

Contd...

Table 9.5–Contd...

Hatching Method	*Merits and Limitations*	*Demerits*
Jars with flowing water supply from above	1. Same merits as in respect of jars with water supply from below.	1. Of complex design and expensive. 2. Need constant vigil and care. 3. If made of transparent plastic, may, with use, become translucent and opaque and hence would need replacement. 4. Can break if made of glass, unless of unbreakable glass which is very expensive. 5. Hatchlings are cut off from direct view.
Circular cemented tank with running water supply	1. Adaptable to large or very large-scale operation. 2. Within limited space, provides effect of some aspects of riverine environment which is the natural habitate of carp. 3. Distribution of oxygen in centrifugal flow of water uniform hence hatching is more effective. 4. Water quality can be controlled. 5. Additional aeration within pool possible. 6. Combines functions of breeding, hatching and larvae rearing tank. 7. Lends itself to modular design. 8. Care-free automatic operation with no special vigil required to watch operation.	1. Requires large quantity of water for effective functioning 2. Monitoring of fertilization and hatching not possible. 3. Defects in leveling in the draining system especially links with hatching tanks are impossible to repair without dismantling the brickwork. 4. Separation of egg shells from hatchlings cumbersome and not very effective. 5. Expensive to install. 6. Not portable.

Contd...

Table 9.5–Contd...

Hatching Method	*Merits and Limitations*	*Demerits*
Rectangular cemented tank with running water	1. Adaptable to large- or very large-scale operation. 2. Provides some aspects of riverine conditions. 3. Distribution of oxygen not uniform because of dead areas at curves or angles, hence, not very effective in hatching operation. 4. Water quality can be controlled. 5. Additional aeration within pool possible but not very effective because of shape.	1. Requires large quantity of water. 2. Monitoring of extent of fertilization and hatching not possible. 3. Defects in leveling in drainage system requires dismantling to repair. 4. Separation of egg shells from hatching not possible. 5. Relative to circular tank less expensive to install. 6. Not portable. 7. Does not lend itself to modular design very effectively because of shape.
Low density polyethylene hatching system–"Dwivedi" hatchery	1. Adaptable to large-scale operation but not to very large-scale operation. 2. Water quality can be controlled. 3. Provides some aspects of riverine conditions. 4. Distribution of oxygen uniform, hence, hatching effective. 5. Additional aeration possible. 6. Extremely easy to repair and mend without dismantling anything. 7. Not expensive to install.	1. The only drawback is that it cannot lend itself to very large-scale operation. 2. Requires vigil and care to operate.

Figure 9.18: Plastic Hatching Jars with Water Supply from Above at the Raipur Carp Hatchery, Noakhali, Bangladesh

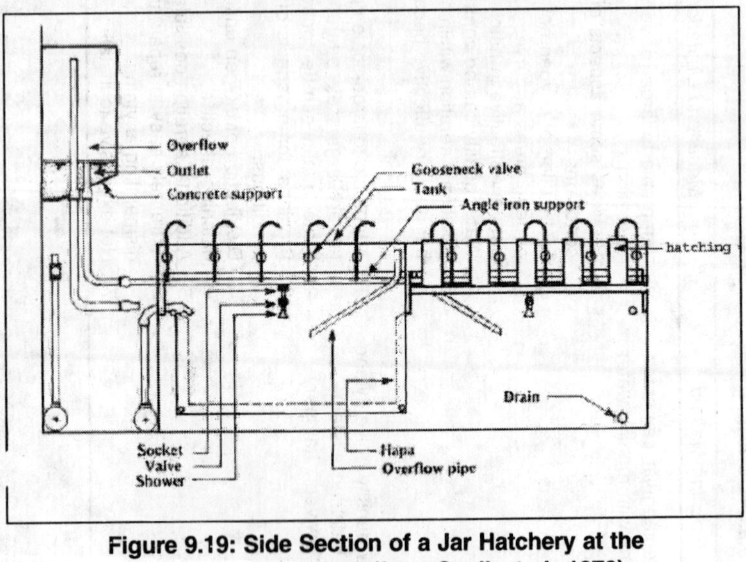

Figure 9.19: Side Section of a Jar Hatchery at the Raipur Carp Hatchery (from Snell *et al.*, 1979)

Figure 9.20: A Cemented Hatchery Tanks Located Directly Below the Hatching Jars at the Raipur Carp Hatchery Jars (from Snell *et al.*, 1979)

Operational Procedures of a Hatchery Proper

Whether the table-top kind or jar-tank kind, the hatchery proper still remains a small-scale unit. For large–scale commercial production of carp seed, a system comprising the ante- or holding tanks and circular tanks, combining the functions of spawning, hatching and fry rearing tank in one, is the best and most effective system so far developed.

The following operational procedures are to be observed for managing these two tank types of systems :

1. Before releasing any broodfish, the ante-tank should be filled with pond water and an antiseptic substrate like potassium permanganate or acriflavin added.
2. To prevent broodfish from jumping out, the ante-tank should be covered with netting having heavy weights (like seine-nets-sinkers).
3. Only filtered, clear, cool, clean and oxygenated water at 27ºC should be used in circular tanks.
4. Its water level and outflow should be controllable by operating turn-down pipes.

5. The rate of flow of water in the circular tanks should be 30-45 liters per minute.
6. There should be a net cover with sinker-like weights for the breeding tank which should be used to cover the tank after the broodfish have been injected and released in it.
7. To safeguard against unexpectedly early breeding after injection (or occasionally even without injection) screens should be put in position.
8. After spawning has taken place, a circular air diffuser should be installed at the base outside the screen and air from compressor of blower bubbled to keep the screen free from eggs.
9. Spawners must be removed from the circular tank after eggs have been completely spawned. They may be given prophylactic treatment in ante-tanks before releasing in a broodfish pond or spent fish pond for possible subsequent maturity.
10. After hatching has occurred, the rate of flow of water through the circular tank should be increased to 45–50 liters per minute and a 6–8 mm meshed nylon net stretched across the tank in a slanting position to collect the discarded egg shells. The nylon net should be retrieved every few minutes to collect cast-off egg shells.
11. If the clogging caused by egg shell bits is not remedied, the central screen fitted into the drain should be carefully replaced by a new one.
12. The hatchling may be left in the circular tanks for four days if a second shift of egg nursing is not to be undertaken. If a second shift is to be undertaken within this period, then the hatchlings should be seined off and removed to the ante-tank.
13. It is essential to periodically brush and hose-wash the screen and keep it completely clean after it has been used one in a breeding tank for few continuous days.

Larval Rearing Tanks

If the breeding tank is not be used as a hatching tank and for further rearing as a larvae rearing tank, then the eggs have to be transferred from breeding tank to the hatching tank, and later the hatchling has to be transferred to larvae rearing tanks before either their packing for transport or for stocking into nurseries for further rearing into fry.

Mention has been made of a modular type of design in which circular breeding tanks may automatically lead to circular hatching tanks. In this design, the latter may also be made to lead to larvae rearing tanks are correctly.

When the fry stage has been reached, *i.e.*, after the yolk has been mostly absorbed and the young fish begin to require exogenous food, the young fish have to be taken out of the hatchery. In some fish farms, the transfer from breeding tank or hatchery jars or funnels to the larvae rearing pond for conditioning before transport is automatically done. For automatic transfer of hatchlings to the rearing pond, the outlets of the breeding tanks have to lead to outdoor earthen nurseries in which *hapas* are installed.

If the conditioning and further rearing of the hatchling is to be done inside the hatchery buildings, then suitable basins for larvae rearing have to be provided. A set of 1.2 m × 1.2 m × 1.2 m basis may be made and hatchling nursed there in for further development and growth.

Physical Facilities

The key facilities required for a commercial hatchery comprise an indoor hatchery and a hatchery pond unit consisting of a number of earthen ponds. Both indoor hatchery and hatchery pond need to be provided with adequate water supply and drainage systems.

Indoor Hatchery

The main function of an indoor hatchery is to produce carp hatchlings under controllable conditions. The facilities in Hungarian hatchery should include a building and a series of concrete cisterns with water supply and drainage facilities. Cistern are used for : (i) holding the spawners, (ii) hatching the eggs after they are fertilised artificially, and (iii) nursing the hatchling for 5-6 days immediately after hatching. Water with a low temperature is required for both holding and hatching cisterns to ensure environments suitable to carps spawn and to minimize the mortalities during egg hatching and hatchling nursing caused by diurnal temperature fluctuation. Ground water with a constant temperature range between 18° C and 25° C would be a most desirable source of supply. Alternatively a circular hatchery may be adopted for producing spawn.

Hatchery Pond

Earth ponds are required for the purpose of (i) growing the hatchling produced from indoor hatcheries to attain fry/fingerlings stage at desirable sizes demanded by markets, and (ii) rearing the brood-stocks of selected varieties to reach full maturation in time for spawning. The proportion of pond area to be used for rearing of fry/fingerlings and broodfish should be arranged in accordance with schedules for carp fry/fingerling production and preparation and selection of brood fish. On average 75 per cent of the total pond area for fry/fingerling production and the rest of the area for brood fish rearing should give a reasonable balance. All ponds can be constructed for both purposes of use.

General Planning in Carp Hatchery Construction

Indoor Hatchery (Hungarian Type)

Cistern Size and Hatchling Production

The size (and the capacity of hatchling production) of a cistern should be determined on the basis of hatchling production targets as well as the standard of technological development in controlling the mortalities of egg hatching and hatchling rearing at each production process.

Example

A targeted hatchling production from an indoor hatchery is 50 million (at a growing stage of 5-6 days after hatchling). The average hatchling rate of artificially fertilised eggs is say, 90 per cent and the average survival rate of hatchlings, 80 per cent. Assuming that the egg hatching capacity of a cistern is 0.15 million m^3, and a total of 10 batches of hatchlings could be produced during the carp breeding season, the size of egg hatchling cisterns would be :

$$50 \times 1/10 \times 100/90 \times 100/80 \times 100/15 = 26.3 \ (m^3)$$

Further assuming that the holding capacity of spawners in a cistern is 15kg/m^3, and the number of eggs is to be released by stripping is 0.1 million/kg of the spawners, the size of a holding cistern required would be :

$$50 \times 1/10 \times 100/90 \times 100/80 \times 10/1 \times 1/15 = 4.6 \ (m^3)$$

Cisterns can be designed in a rectangular shape with a depth of about 1.0 m. Each cistern requires water supply and drainage systems and its water level needs to be regulated as needed. A hatchling catch basin to facilitate the collection of hatchling catch basin to facilitate the collection of hatchlings from the cisterns should be constructed at one side of each hatchling cistern. The volume of water required for the operations of an indoor hatchery could be calculated at 1.0-1.5 litre/sec for each million of hatchling production. As an improvement over stripping and fertilisation of eggs, one or two circular spawning tanks could be added inside the hatchery building.

Hatching Net

Although other measures are known to be successful, on the basis of availability and costs, funnel nets are recommended to suspend egg hatchling in the cistern. These nets have a number of advantage (i) they make, it easy to remove the egg shells (after the hatchlings have hatched out) from the cistern without polluting the water, (ii) they are easy to handle, and (iii) they are easy and cheap to make from local materials. Each net should have a capacity of 0.3-0.4 m^3 and should be capable of holding 50,000-60,000 fertilised eggs, which would be a size convenient for a hatchery man to handle. The Mesh size of the net should be about 0.2 cm^2 or 32 meshes/ square inch/ which would be maximum size to hold the fertilised eggs throughout the hatching period, but a minimum size to allow the newly hatched hatchling to drop from the net to the cistern bottoms. The delicate hatching should be intensively nursed in the cistern for 5-6 days without handling. Each funnel net is connected to a water supply pipe and to flow within a net would be regulated by a funnel. The fertilised eggs should be carried by the flow to circulate gently within the net.

Hatchery Pond

Pond Size and Fry/Fingerlings Production

The production of carp fry fingerlings from a pond depends primarily on pond productivity. Pond productivity varies with the conditions of the environment, such as the range of temperature, intensity of sun light, depth of water, and richness of soils, as well as the techniques of pond management, such as the application of fertilizers, control of water quality and manipulation of fish population.

Example

Annual requirement of 5 cm fingerlings (weighing an average 1.5 g or about 700 fingerlings/kg) is 12 million. Assuming that the standing crop of carp fry/fingerlings produced from a hatchery ponds is 1.2 ton/ha and a minimum of two crops can be produced per carp breeding season, the area requirement for hatchery pond can be calculated by :

$$12 \times 10^6 + 1/2 \times 1/700 \times 1.2 \times 10^3 = 7.3 \text{ (ha)}$$

Further assuming that the average survival rate of carp fingerlings raised from hatchling state is 30 per cent then the total carp hatchling required for stocking the hatchery pond would be :

$$12 \times 10^6 + 100/30 = 40 \times 10^6$$

The project report may be prepared according to the following standard headlines:

1. Introduction
2. Details of the site
   (*i*) Location
   (*ii*) Topography
   (*iii*) Nearness to the main road, supply of electricity water etc.
   (*iv*) Flood and drainage conditions
3. Soil qualities of the site
4. Water table conditions
5. Water requirement
6. Estimates for all works
7. Designs
8. Plans
9. Annexures-including the laboratory test results, drawings etc.

Table 9.6: Items of Work Necessary and Through Cost Estimates for the Establishment of a 10 ha. Size Fish Seed Farm-cum-Hatchery (Based upon rates for the year 2010-2011)

Sl.No.	*Item of Work*	*(Rs. in Lakhs)*	*Remarks*
1.	Civil works		
1.1	Site preparation earth work etc.	9.50	
1.2	Hatchery roads including approach road upto 1 km length	3.50	
1.3	Water gates	9.00	
1.4	Water supply	6.00	
1.5	Hatchery Building	3.00	
1.6	Building for office, staff quarters, store etc. including electricity supply fencing and gates	14.00	Precise estimate will depend upon actual site conditions
1.7	Protection of pond banks/pitching and site finishing etc.	2.00	
1.8	Engineering charges @ 10 per cent of cost of civil works	4.70	
2.	Vehicles and Equipment (as per Table 9.7)	5.34	
3.	Contingencies including escalation in Price L/S Total cost per 10 ha size fish seed farm project excluding cost of land	64.00	
	Total cost per project	**74.00**	

Table 9.7: Requirement of Stores/Equipment/Vehicle etc. for a 10 ha Size Seed Farm-cum-Hatchery 1

Item	*Nos.*	*Rupees*
Light duty truck 2/	1	40,000
Jeep with trailer 3/	1	1,25,000
Portable water pumps	3	90,000
Hatching nets 5/	40	8,000
Office furnitures		20,000
Nets (Breeding)	6	4,800
Drag nets	1	2,500

Contd...

Table 9.7–Contd...

Item	*Nos.*	*Rupees*
Gill Net	1	3,500
Kit for hypohysation 6/	10	8,000
Hand centrifuge 7/	3	25,000
Aerators 8/	5	75,000
Electric Generator	1	1,25,000
Total Cost		5,30,000

1/ Based on the 10 ha. hatchery model

2/ Three wheel tempo Auto

3/ Four wheel diesel jeep

4/ 5. H.P. low lift turbine pumps of ½ cusec ca @ 30,000 each

5/ Each hatching net @ Rs. 200

6/ One kit @ Rs. 800/–each kit will contain 3 pairs of syringe, different size injection needless and 12 glass vials/ tube and tissue grinder etc.

7/ For preparing pituitary gland extract.

8/ Aerators for oxygeration and accelerated growth of fish seed and brooders in the earthen ponds.

9/ Electric generator (including cost of generator house etc) to ensure continuous power supply and hence water supply etc. in the Indoor hatchery building and to maintain required water level in the ponds.

A Typical Fish Farm

Dependable sources of water supply to ponds and soil which will retain water are of fundamental importance in the selection of a site for pond construction. Low-lying water-logged or marshy areas are suited to excavation into ponds, but the soil, if peaty, would reduce the productivity of waters. A clay-loamy soil is perhaps the best from the point of view of water retention and fertility.

When it is intended to rear carps in new ponds, the requirements for nursing, rearing, and stocking fish in different types of ponds should be remembered. A small farm, 10 acres in extent, should have the area divided as follows for different purpose :

Nurseries	½ acre
Rearing ponds	2 acres
Stocking ponds	7½ acres

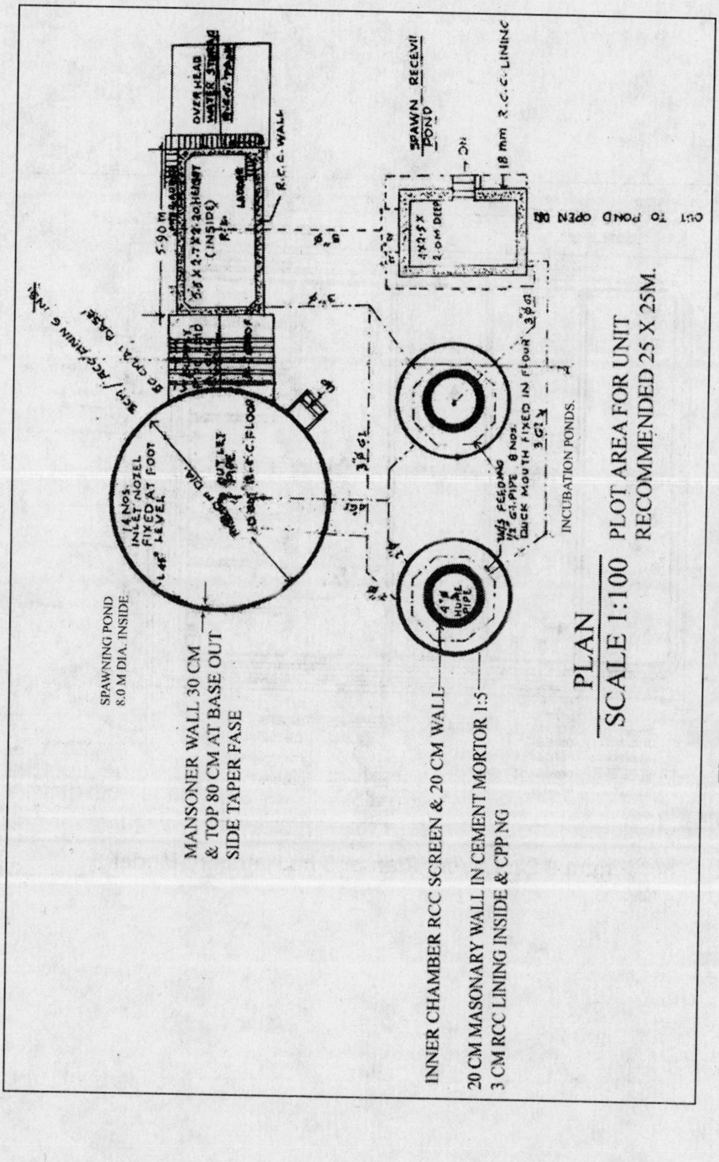

Figure 9.21: Circular Fish Hatchery

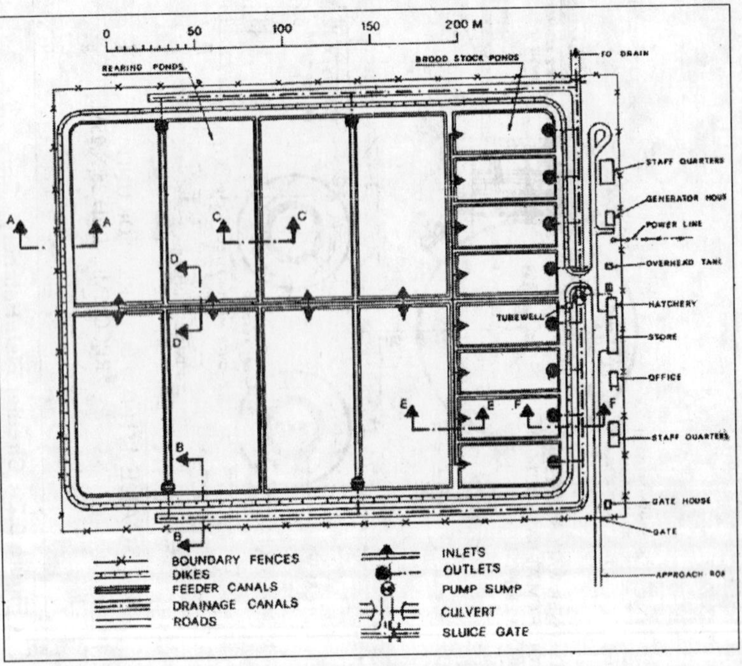

Figure 9.22: Layout Plan of 5 ha Hatchery Model

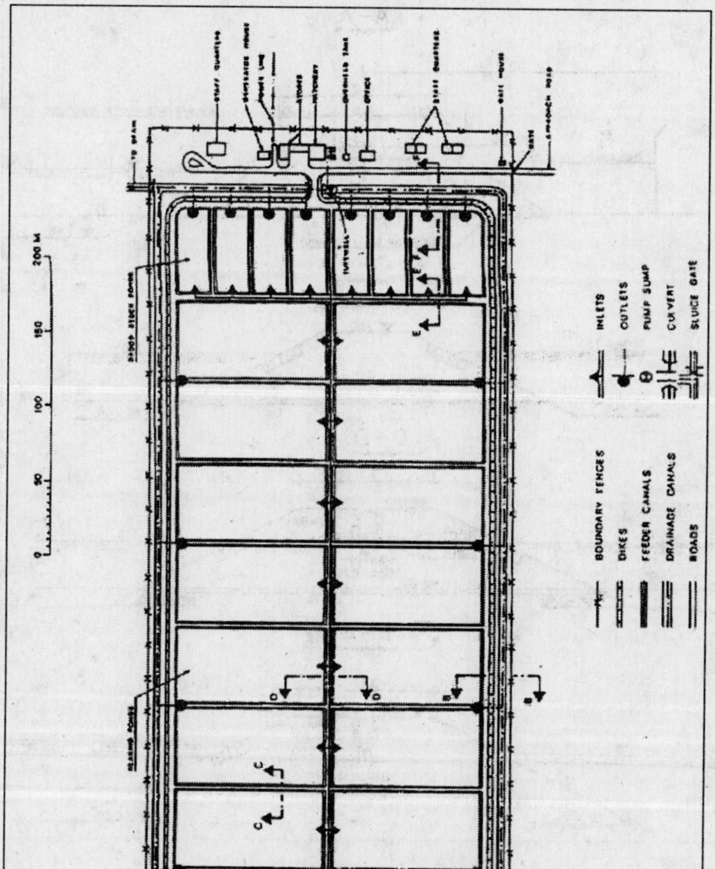

Figure 9.23: Layout Plan of 10 ha Hatchery Model

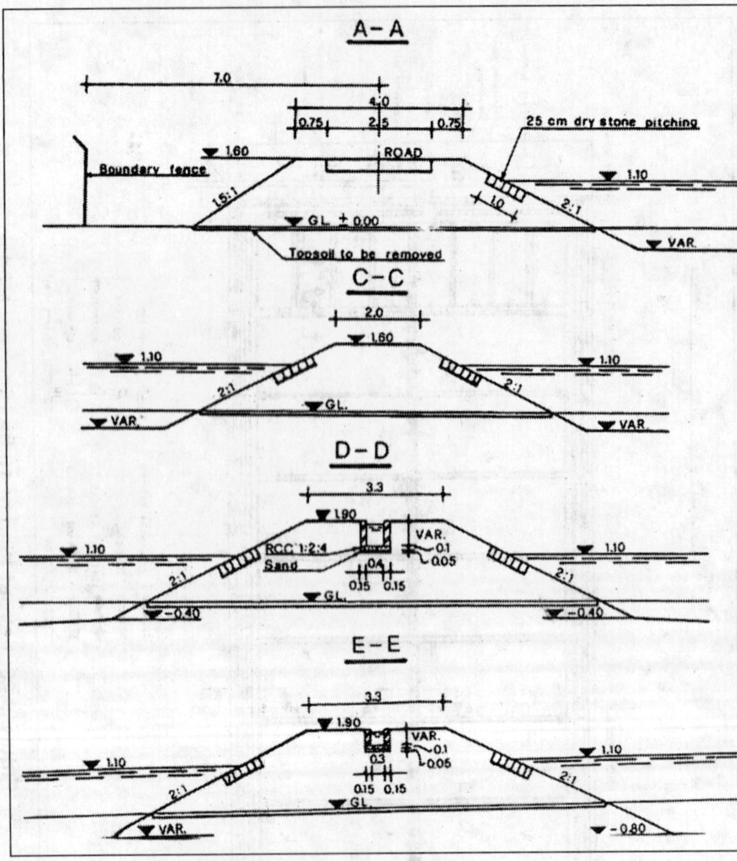

Figure 9.24: Cross Sections of Dikes

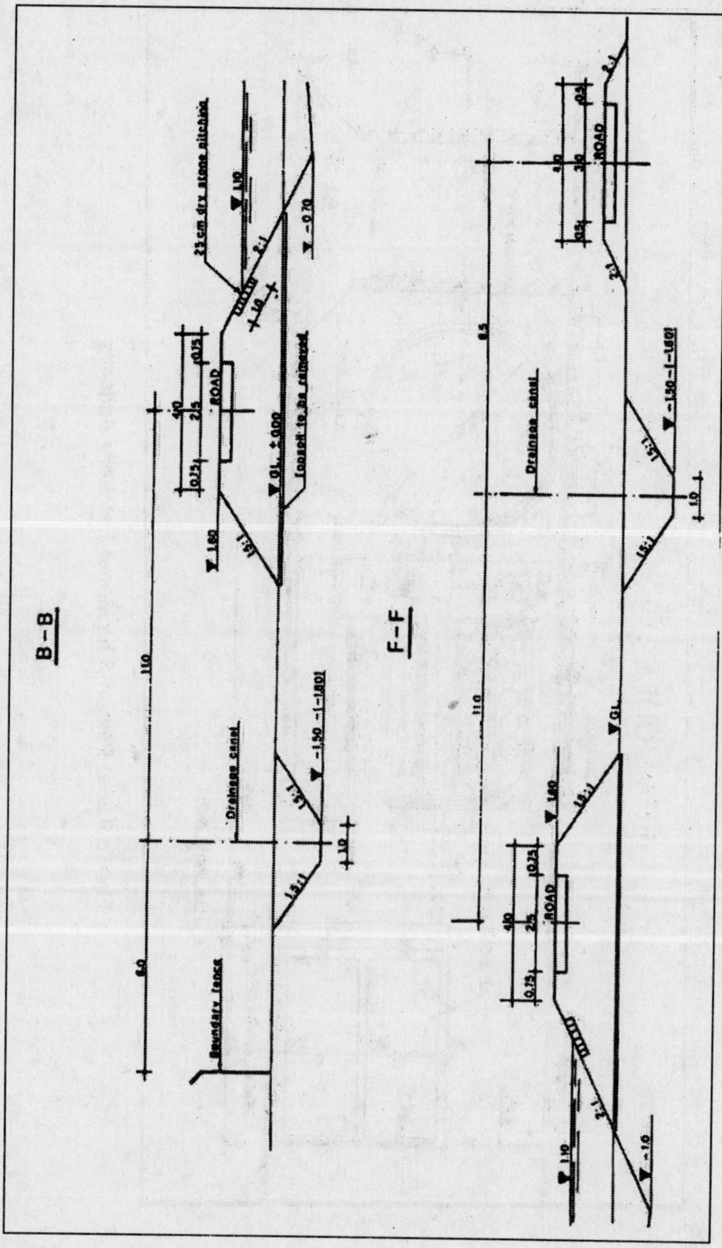

Figure 9.25: Cross Sections of Dikes

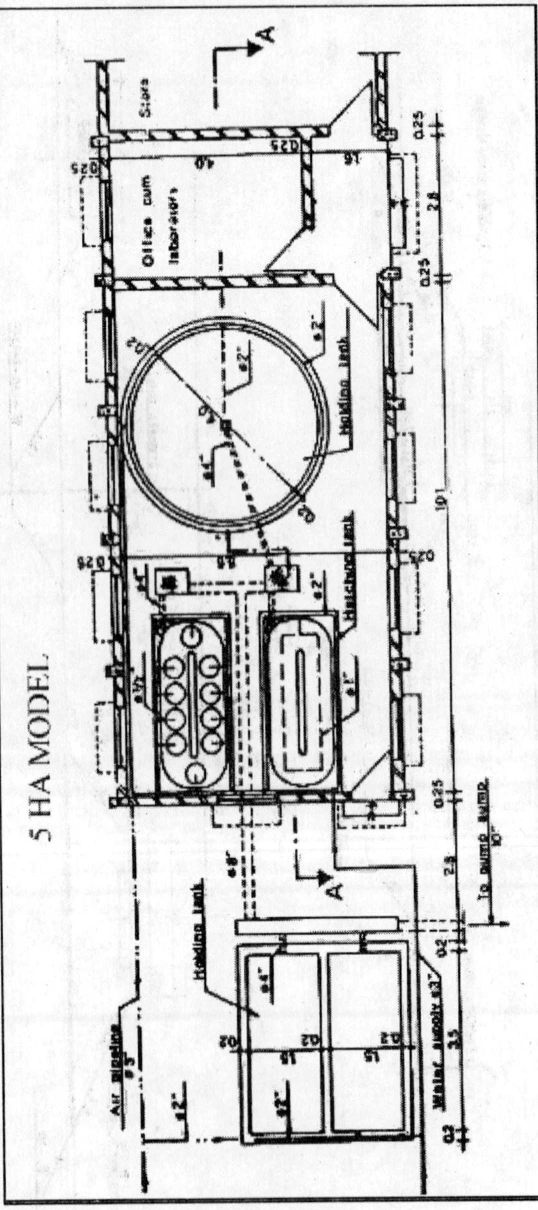

Figure 9.26a: Plan of 5 ha Model Hatchery Building

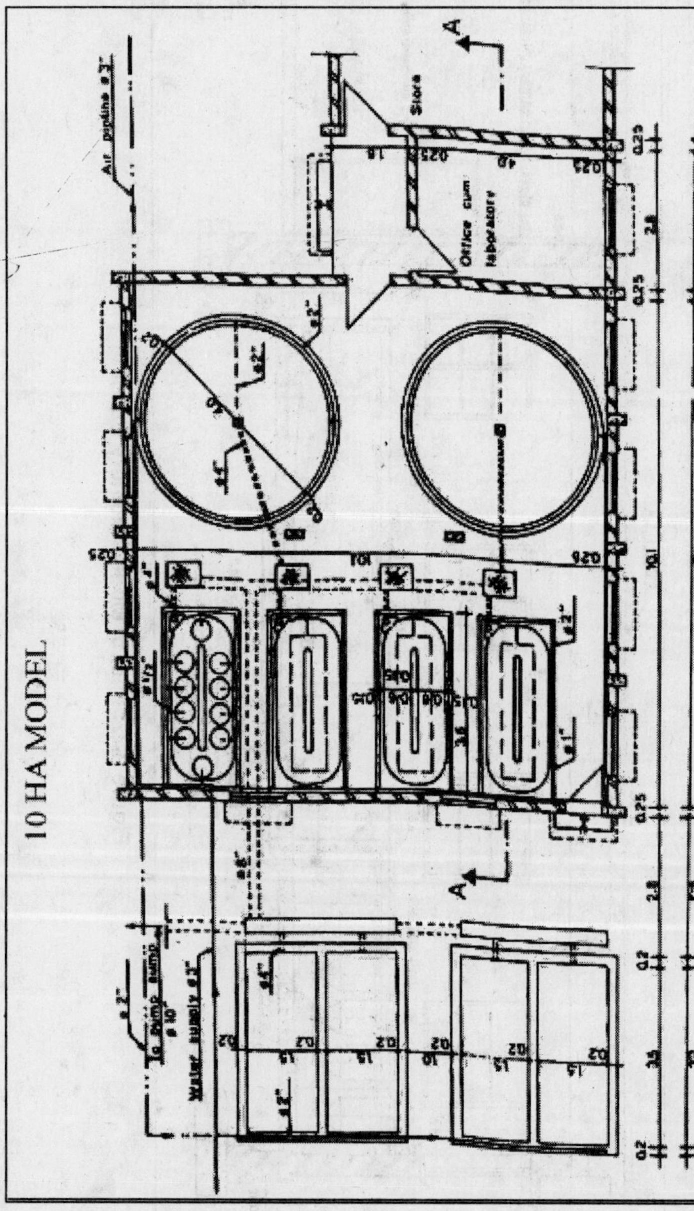

Figure 9.26b: Plan of 10 ha Model Hatchery Building

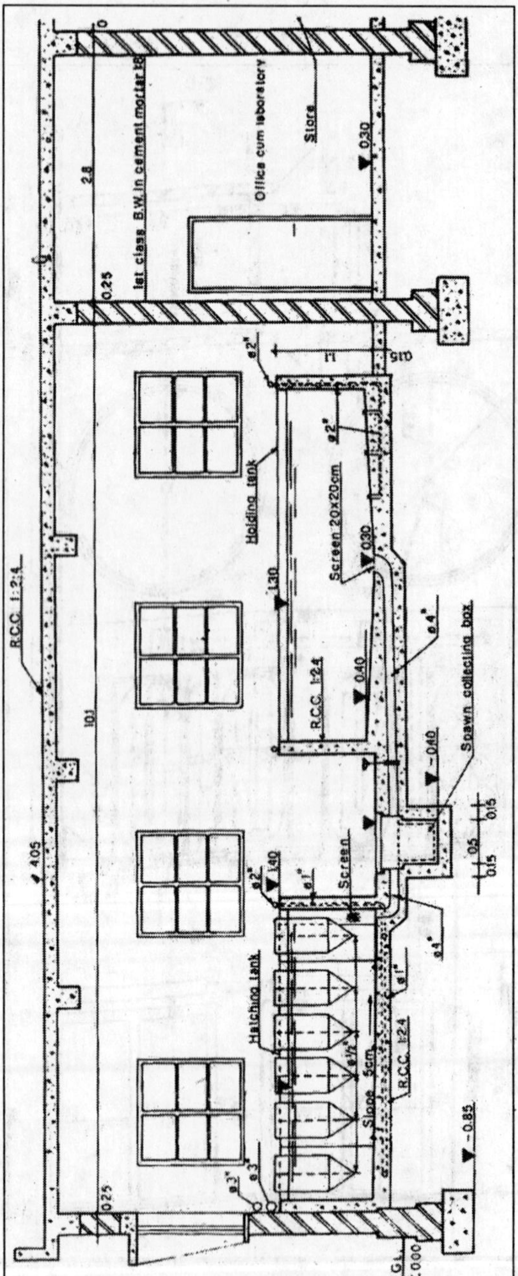

Figure 9.27: Plan of Model Hatchery Building

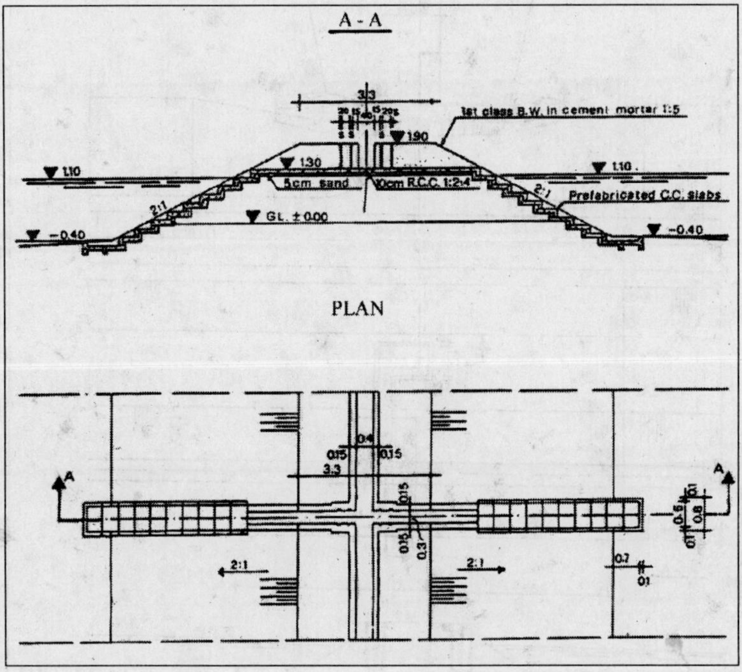

Figure 9.28: Inlet Plan and Cross Section

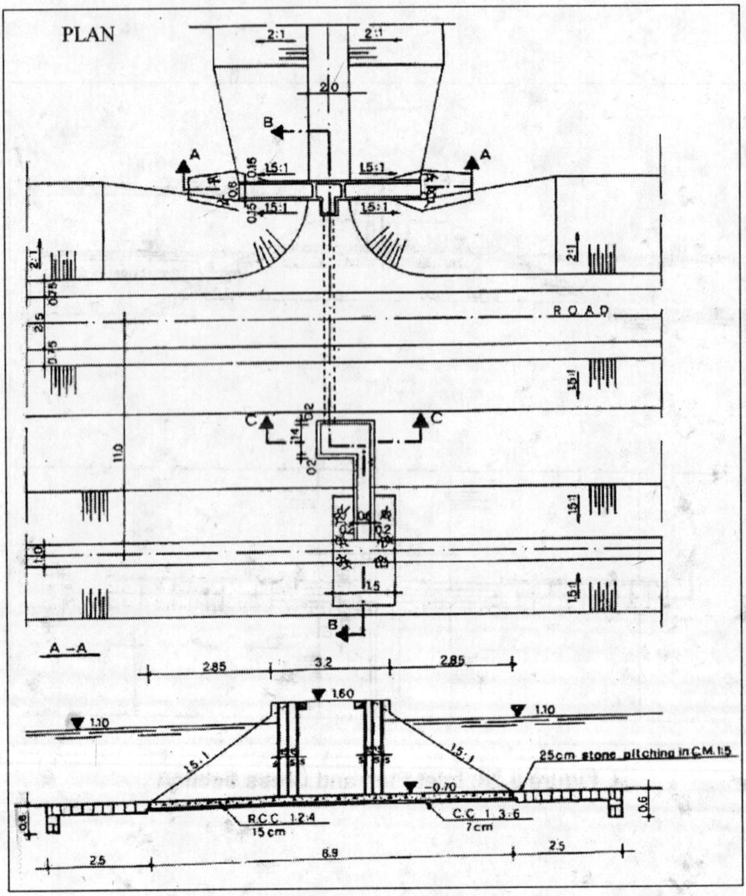

Figure 9.29: Outlet Plan and Cross Section

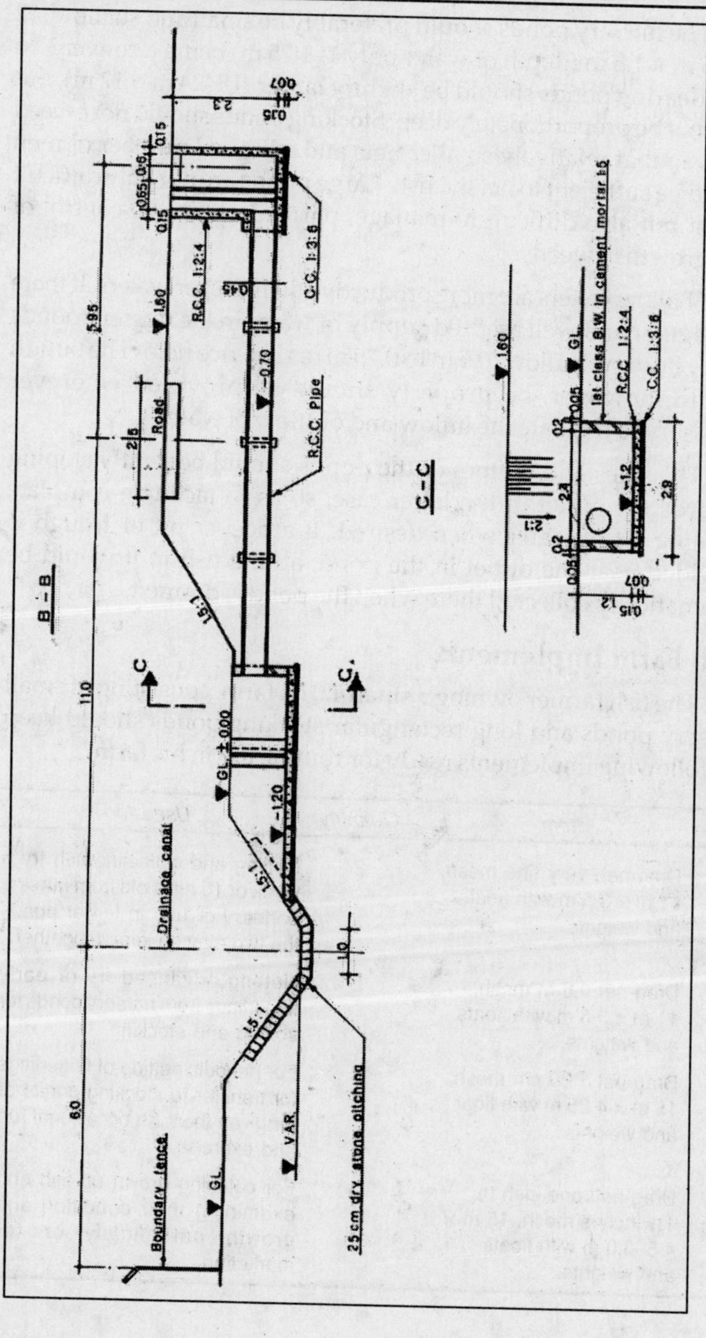

Figure 9.30: Outlet Cross Sections

The nursery ponds should preferably be small and shallow, 12 m × 6 m × 1.5 m (depth of water only 1–1.25 m) being a convenient size. Rearing ponds should be slightly larger (18–24 m × 12 m), but need not be proportionately deep. Stocking ponds should not exceed 30 m, so that relatively smaller nets and a limited number of men would be sufficient to net the fish. Large ponds are not only difficult to fish but also difficult to manage, particularly in the control of overgrowth in weeds.

Shallow waters are more productive than deeper waters. If there is a regular and well assured supply of water in the nursery ponds it may be very shallow (0.5 m to 0.75 m) as in a rice field. The bundh should, however, be properly should be provided wherever necessary to regulate the inflow and outflow of water.

The sides and bottom of the ponds should be gently sloping towards the outlet in the latter case, so as to facilitate complete draining off of water when desired. If a deeper pit or trough is provided near the outlet in the pond, all the fish in it would be automatically collected there when the pond is drained.

Fish Farm Implements

The fish farmer owning a small 4.0 ha farm, consisting of small nursery ponds and long rectangular stocking ponds should keep the following implements ready for routine use in his farm.

Sl.No.	*Item*	*Quantity*	*Used for*
1.	Drag-net, very fine mesh, 11 m × 3.5 m with floats and weights.	2	Netting and collecting fish fry a week or 10 days old from nursery ponds. For use in larger ponds the two may be joined together.
2.	Drag-net 0.6 m mesh, 11 m × 3.5 m with floats and weights.	2	Netting advanced fry or early fingerlings from nursery ponds for sorting and stocking.
3.	Drag-net 1.25 cm mesh; 15 m × 4.25 m with float and weights.	3	For periodic netting of fingerlings for transfer to stocking ponds or for given them an occasional run and exercise.
4.	Drag-net, one inch to 1½ inches mesh, 15 m × 4.5–6.0 m with floats and weights.	3	For catching grown up fish and examining their condition and growth periodically or for marketing.

Sl.No.	Item	Quantity	Used for
5.	Miniature cloth tank or hapa, 2 × 1 × 1 or 1.5 m with strong tape at all corners, top and bottom.	2	For temporary storage of fish after netting from pond before sales, necessary also for sorting fingerlings of major carps from others.
6.	Wide mouthed earthen pots or hundies about 18–22 liters capacity (tins preferable, but costly).	6	For temporary storage of live fish fry, fingerlings or adults, after netting and for transport from one place to another.
7.	Buckets.	6	For carrying fish seed, manure etc. from place to place.
8.	Baskets.	1 dozen	For carrying sand, silt, manure, etc.
9.	Spades.	2	For trimming embankments, desilting etc.
10	Jack knife.	4	For trimming weeds on the embankment and in the marginal areas of ponds, etc.
11.	Bamboo poles, 2.5 m to 3 m long.	1 dozen	For fixing the *hapas* in ponds and for other miscellaneous use.
12.	Spilt bamboo screen; 3 m × 1.25 m.	2	For use as miniature conditioning tank or live pen in the pond for fish to be kept alive after catching but before marketing.
13.	Weeder or bottom rake (a wooden or metal beam) 2 to 2.5 m long, with strong bamboo or metal spikes fixed in all directions and with strong rope tied at either end for dragging.	1	De-weeding deeper areas of ponds by repeated dragging in all directions and removing the weeds caught among the spikes every time the weeder is hauled, and disturbing the bottom mud to remove poisonous gases accumulating therein.
14.	Straw rope, kloose, thick, about 100 feet long.	1	Dragging the surface water to remove floating weeds (*Pistia, Lemna, Azolla, Wolffia*), thick algal scums, etc., which often occur in great abundance in ponds.
15	Small boat or float.	1	De-weeding purposes, particularly when the pond is deep and broad and abounds in rooted weeds like water lilies which have to be plucked out.

Other miscellaneous items may also be required occasionally. An extra stocking of wire-netting for protecting the inlets and outlets of ponds, if any, a balance with weights for weighing fish, a scale (in inches and centimeters) for measuring their length, etc. will be necessary. A proper maintenance of the farm will require the full-time services of at least two persons who could also profitable be engaged in vegetable farming on banks and bunds of fish ponds.

The pond should be numbered serially, and the details regarding their dimensions, etc., entered in a maintenance register, one page for each pond. The physical characters of the water in the ponds, and the forms of life observed in the ponds including plants, insects, fishes and other animals-big, small and immature, the farming operations such as manuring stocking, netting, marketing, size of fish caught their number etc. In fact, every little operation or observation, including behaviour of fish, should be recorded in the book then and there to facilitate institution of comparisons of the different conditions at different times with a view to improving the farming methods.

Chapter 10

Methods of Analysis for Waters, Organic Matter and Pond Bottom Soils

Winkler Method for Determination of Dissolved Oxygen

This method of oxygen analysis determines the free dissolved O_2 in water. It is based on the following reactions:

1. $MnSO_4 + 2\,NaOH \rightarrow Mn(OH)_2 + Na_2SO_4$
2. $Mn(OH)_2 + O \rightarrow H_2MnO_3$ (brown precipitate)
3. $H_2MnO_3 + 2\,H_2SO_4 + 2\,KI \rightarrow MnSO_4 + HOH + K_2SO_4 + I_2$

The released I_2 gives the solution a yellow to brownish colour, and the intensity of the colour is proportional to the O_2 content of the original water sample. The amount of I_2 is determined by titration with sodium thiosulphate :

4. $I_2 + 2N_2S_2O_3 \rightarrow Na_2S_4O_6 + 2\,NaI$

The end point is determined by the use of starch indicator, which gives a blue colour with free I_2 and no colour with NaI. The disappearance of the blue colour marks the end of the reaction.

If the titrated solution is allowed to stand for several minute, subsequently the blue colour returns due to absorption of additional

O_2 from the air and the release of I_2 from the HI present in the acid solution.

5. $4\,HI + O_2 \rightarrow 2\,I_2 + 2\,H_2O$

Also, presence of Fe (ic) or NO_2 may cause subsequent release of free I_2 from KI or NaI. Consequently, the sample should not be again titrated after the original end-point has been reached.

Meterials Interfering with the Accuracy of the Method

Certain materials when present in waters interface with the accuracy of the Winkler method.

Turbidity

Muddy water and those with heavy suspended organic matter should be cleared as follows :

To a 1-liter water sample, add 10 ml of 10 per cent potassium aluminum sulphate.

Then add 1 ml 35 per cent NaOH. Stopper and rotate. Allow the precipitate to settle and analyze the clear supernatant liquid. (Avoid excess of NaOH because at pH above 11, organic mater absorbs O_2 rapidly)

For activated sludges and heavy organic matter, clear as follows:

Add 10 ml copper-sulphamic acid to 1-liter bottle. Add water sample by siphoning. Mix, and allow to settle.

Copper-sulphamic Acid

Dissolve 16 gm sulphamic acid in 200 ml distilled water. Dissolve 25 gm $CuSO_4 \cdot 5H_2O$ in 250 ml water. Mix and add 12 ml glacial acetic acid.

Materials Causing High Results

Nitrites and ferric iron tend to cause high results because they react with KI to release I_2.

Nitrites are present in most water containing fish. For more accurate results, the sulphamic acid modification or the potassium permanganate modification may be used. The reactions are as follows:

$$HNO_2 + NH_2SO_2OH \rightarrow H_2SO_4 + N_2 + H_2O$$

$$2\,KMnO_4 + 5\,NaNO_2 + 3H_2SO_4 = K_2SO_4 + 2\,MnSO_4 + 5NaNO_3 + 3H_2O$$

Ferric Iron

Except at concentrations in excess of 5 ppm, ferric iron causes little error. High concentrations of ferric iron are normally present only in strongly acid waters. The addition of 1 ml of 40 per cent KF. $2H_2O$ is sufficient to prevent interference up to 200 ppm Fe^{+++}. It should be added prior to starting the standard Winkler determination. This treatment is principally necessary in analyzing water from streams receiving acid mine wastes.

Materials Causing Low Results

Ferrous iron, sulphites and organic matter cause low results by removing the liberate I_2 from the treated sample. Organic matter may absorb I_2 especially at high alkalinities, and also cause loss of dissolved O_2 prior to the analysis.

Ferrous Iron

A concentration of 1 ppm, Fe^{++} causes the loss of 0.14 ppm O_2. Since ferrous iron is principally present in water very low or lacking is dissolved oxygen (DO), it seldom requires removal in well oxygenated waters. When present in excess of 1 ppm, the water sample should be pretreated with the potassium permanganate modification :

$$10\,FeSO_4 + 2\,KMnO_4 + 8\,H_2SO_4 \rightarrow 5\,Fe_2(SO_4)_3 + 2\,MnSO_4 + K_2SO_4 + 8\,H_2O$$

Sulphites and Organic Matter

These materials found in paper mill effluents make analysis of oxygen difficult. The alkaline hypochlorite modification has been proposed (see APHA Standard Methods) but the reagents are difficult to prepare and must be restandardized weekly. In river waters receiving such effluents, the sulphamic acid modification gives reasonably accurate results. When organic matter is present in sufficient amount to make the water murky, the water should be treated with Cu-sulphamic acid as explained above. The analysis of the clear supernatant liquid should then proceed promptly and rapidly.

Preliminary Tests to Determine Presence of Interfering Materials

Reagents

1. *5 per cent KI* : Dissolve 5 gm KI in 100 ml distilled water.

2. *HCI*: Concentrated.
3. *Iodine stock*: Dissolve 6 gm KI + 4 gm I_2 in 100 ml distilled water.
4. *Dilute I_2 solution*: Add 1 drop stock Iodine to 100 ml distilled water. Store in brown glass-stoppered bottle not over 3 days.
5. *Potassium thiocyanate 5 per cent*: Dissolve 5 gm in 100 ml distilled water.
6. *Potassium ferrocyanide 5 per cent*: Aqueous solution, as above.
7. *Potassium ferricyanide 5 per cent*: Aqueous solution, as above.
8. *Sulphanilic acid solution*: Dilute 30 ml glacial acetic acid to 100 ml with distilled water. Add 1 gm sulphanilic acid
9. *Alpha-Naphtylamine acetate*: Dilute 30 ml glacial acetic acid to 100 ml with distilled water. Add 0.5 gm alpha-naphtylamine, shake and filter through cotton. Store in a brown bottle in the dark.

Procedure

1. Collect about 50 ml of water sample in an open beaker.
2. Filter, if necessary, through filter paper or cotton.
3. To a 10 ml water sample, add 1 ml of 5 per cent KI and 1 ml concentrated HCl. Mix. A brown to yellow colour indicates presence of materials causing high results. See Section A following.
4. To a 10 ml sample add 2 drops starch and then drop by drop add dilute I_2, shaking after each drop. Record the number of drops required for a permanent blue colour.

In another flask, place 10 ml distilled water and add 2 drops starch solution. Add drop by drop of dilute I_2 solution shaking after each drop. Record the number of drops required for a permanent blue colour.

If the drops required for the water sample are less than that for the distilled water, this indicates the presence of materials absorbing I_2 and causing low results *e.g.*, ferrous iron, sulphites, organic matter and /or similar substances. See Section B following.

A. Materials Causing High Results (Releasing I_2)

a. Nitrites

Test : Acidify a 50 ml water sample with a drop of HCI. Add 2 ml of sulphanilic acid and 2 ml naphthylamine hydrochloride. Nitrites cause a pink colouration.

b. Ferric Iron

Test: Acidify a 10 ml water sample with 2 drops conc. HCl. Add 0.5 ml 5 per cent K thiocyanate. A pink to red colour indicates ferric ion; 10 ppm Fe^{+++} gives a very slight pink and 15 ppm a light red.

Test: To a neutral or only slightly acid water sample add several drops K ferrocyanide. The colour varies from green to blue with increasing amounts of Fe^{+++}; 1 ppm gives a faint blue-green on standing, 5 ppm gives a light blue within several minutes and 10 ppm gives a deep blue. The reaction is:

$$4\,FeCl_3 + 3\,K_4\,Fe(CN)_6 \rightarrow \underset{\text{prussian blue}}{Fe_4\,(FeC_6N_6)} + 12\,KCl.$$

This test in sensitive to 0.002 mg ferric ion. Ferrous iron may gives a white precipitate.

B. Materials Causing Low Results (Absorbing I_2)

a. Ferrous Iron

Test: To a 10 ml water sample add a few drops of potassium ferricyanide solution. A blue precipitate indicates the presence of ferrous iron (Fe^{++}). The reaction is :

$$3\,FeSO_4 + 2\,K_3\,Fe(CN)_6 \rightarrow \underset{\text{Turnhill blue}}{Fe_3(FeC_6N_6)} + 3\,K_2SO_4$$

This blue precipitate is insoluble in HCl. A solution of 1 ppm Fe^{++} gives a light green colour, 10 ppm a blue-green colour and 50 ppm gives a deep blue. Ferric iron may give a brown colour.

Modifications of the Winkler Method

For exceptional waters containing considerable amounts of ferrous iron, the Rideal-Stewart modification of the Winkler method may be used. This involves oxidation of the ferrous iron to ferric iron with potassium permanganate.

For most waters, nitrite (NO_2^-) is the principal interfering salt. Sulphamic acid pretreatment reduces the error from the this interference :

$$NH_2SO_2OH + HNO_2 \rightarrow H_2SO_4 + N_2 + H_2O$$

Since the sulphamic acid is stable in sulphuric acid and easy to use, this modification is used in all waters containing fish.

Sodium azide (NaN_3) may also be used to eliminate interference by nitrites :

$$NaN_3 + H_2SO_4 + HNO_2 \rightarrow N_2 + H_2O + N_2O + NaHSO_4$$

This method is not given although it is as effective as the sulphamic acid modification.

The Iodine Difference Method of Ohle may be used in problem waters or when great accuracy is desired. This corrects both for the presence of materials causing high results by releasing iodine from KI and causing low results by absorption of free iodine.

Reagents

For Rideal-Stewart Modification

1. *Potassium permanganate solution*: Dissolve 3.16 gm potassium permanganate in distilled water and dilute to 500 ml. Transfer to a clean brown bottle and keep in the dark.
2. *Potassium oxalate solution*: Dissolve 10 gm of potassium oxalate in water and dilute to 500 ml.

For Sulphamic Acid Pretreatment of Water

1. *Sulphamic acid solution*: Add 20 ml concentrated sulphuric acid to 80 ml water. Add 4 gm sulphamic acid and shake until dissolved. Store in brown bottle with glass stopper.

For Standard Winkler Method

1. *Manganous sulphate solution*: Dissolve 240 gm of manganous sulphate in water and dilute to 500 ml.
2. *Alkaline potassium iodide solution*: Dissolve 250 gm of sodium hydroxide and 75 gm of potassium iodide in water and

dilute to 500 ml. Use a flask with a rubber stopper for storage.

3. *Standard 0.025 N sodium thiosulphate solution*: In reaction 4, page 257. 2 molecules of sodium thiosulphate reacted with 2 iodide equivalents. Therefore, an N/1 solution of sodium thiosulphate contains the molecular weight of this compound in gm per liter of solution. The salt used in $Na_2S_2O_3 . 5H_2O$, with a molecular weight of 248.19. A 0.025 N solution then contains.

   $$248.19 \times 0.025 = 6.2048 \text{ gm per liter of solution.}$$

   Weigh accurately 6.205 gm of chemically pure recrystallized sodium thiosulphate. Place in a 1 – liter glass-stoppered volumetric flask and dilute to 1 liter with freshly boiled and cooled distilled water. Add several drops chloroform as a preservative, lable, and stopper tightly. Keep this reagent in the dark.

   Each ml of 0.025 N sodium thiosulphate is equivalent to 0.2 mg of oxygen or to 0.1400 ml of oxygen at 20°C and 760 mm pressure.

   As this reagent is not permanent, it should be restandardized at 30 day intervals against a 0.025 N solution of potassium dichromate.

4. *Standard 0.025 N potassium dichromate solution*: This salt makes a permanent standard and should be prepared carefully. It reacts with KI in acid solution to liberate a definite amount of I_2 as follows :

   $$6\,KI + K_2Cr_2O_7 + 14\,HCl \rightarrow 8KCl + CrCl_3 + 7H_2O + 3I_2$$

   The liberate iodine is then titrated with the thiosulphate solution. Since in the above reaction one molecule of $K_2Cr_2O_7$ reacts with 6 iodine equivalents, a normal solution of potassium dichromate equals the molecular weight in grams divided by 6. The $K_2Cr_2O_7$ contains no water of crystallization and has molecular weight of 294.2; a 0.025 N solution of this salt then contains :

   $$\frac{294.2}{6} \times 0.025 = 1.2258 \text{ gm per liter}$$

Weigh accurately 0.6129 gm of pure crystalline $K_2Cr_2O_7$ and dilute to 500 ml in a glass-stoppered volumetric flask with freshly boiled and cooled distilled water.

5. *25 per cent potassium iodide solution*: Dissolve 25 gm KI in 100 ml freshly boiled and cooled distilled water.
6. *Starch indicator*: Add 2 gm soluble starch to 100 ml water. Heat until transparent. Add 0.5 ml formalin for preservative.

Or use:

Modified Starch Glycerite

1. Add 3 gm powdered soluble starch to 100 ml glycerin.
2. Heat to 190° C.
3. Cool to room temperature. If solution gels with age, reheat to 190° C and add glycerin.

Starch indicator is stable for about 30 days, while to modified starch glycerite keep indefinitely.

Standardization of Approximately 0.025 N Sodium Thiosulphate

1. Place in a beaker 100 m˙of freshly boiled and cooled distilled water.
2. Add 3 ml of a 25 per cent solution of KI.
3. Add from a burette 50 ml of the standard 0.025 N potassium dichromate.
4. Add 10 ml concentrated HCl.
5. Titrate the liberated I_2 with the standard 0.025 N sodium thiosulphate solution. When the yellow colour has almost disappeared, add about 2 ml starch indicator and continue the titration until the blue colour just disappears. The final colour may be a light green due to chromic salts.
6. The factor for the 0.025 N thiosulphate equals;

$$\frac{\text{ml standard dichromate}}{\text{ml standard thiosulphate}} = \frac{50}{x}$$

For example, if the titration requires 48.8 ml of approximately 0.025 N thiosulphate, the factors

$$= \frac{50}{48.8} = 1.024$$

Collection of Sample

Collect the sample in a narrow-necked glass stoppered bottle of 250 to 300 ml capacity (standard BOD bottle) by means of an apparatus designed to avoid the entrapment or absorption of any oxygen from the atmosphere. Note the temperature.

Determination of Dissolved Oxygen

Removal of Ferrous Iron and Nitrites

Follow Modification I only, if iron or other interfering materials prevent the use of Modification II.

Modification I : Rideal-Stewart Modification

1. Remove the stopper from the bottle and add 0.7 ml of concentrated sulphuric acid and then 1 ml of the potassium permanganate solution. Introduce these and all other reagents by pipette under the surface of the liquid. Insert the stopper and mix by inverting the bottle several times.
2. If a noticeable excess of potassium permanganate is not present after 5 minutes, again add 1 ml of the permanganate solution; if this is still insufficient, use a stronger permanganate solution, but avoid a large excess.
3. After 20 minutes destroy the excess permanganate by adding 1 ml of the potassium oxalate solution, re-stopper the bottle at once, and mix its contents. Let stand several minutes. If the pink colours persists, add an additional 1 ml of potassium oxalate, re-stopper, and mix. Repeat until solution is free of pink colour, but avoid adding excess oxalate as this causes low results (Continue with Standard Winkler Method.)

Modification II : Sulphamic Acid Modification

Use this modification for most pond waters.

1. Remove the stopper and add 1 ml of the sulphamic acid solution. Replace stopper and mix by inverting bottle. After 30 seconds, continue with the Standard Winkler Method of analysis.

Standard Winkler Method

1. Add 1 ml of the manganous sulphate solution and 3 ml of the alkaline potassium iodide solution. Mix by rapid inversion for 30 seconds. Allow the precipitate to settle. Re-shake and allow to settle (For sea water, alternately shake and allow to settle for a 10 minutes period).
2. Add 1 ml of concentrated sulphuric acid mix by shaking. If water is high in iron or organic matter, continue promptly with steps 3 and 4.
3. Measure 200 ml of the contents of the bottle into a graduated cylinder or pipette. Discard the remainder of the sample and carefully pour the measured sample back into the sample bottle or into a Erlenmeyer flask with the minimum of aeration.
4. Titrate with 0.025 N thiosulphate until the sample becomes a faint Yellow. Add a few drops of the starch indicator and continue the titration until the blue colour disappears.
5. Report results as parts per million. Each ml of standard 0.025 N sodium thiosulphate equals 1 ppm O_2. For example, if the titration required 4.5 ml of 0.025 N thiosulphate, the sample contained 4.5 ppm O_2.

If the thiosulphate had a factor of 1.024, each ml = 1.024 ppm O_2 and the above sample contained:

$$4.5 \times 1.024 = 4.6 \text{ ppm } O_2$$

Modified Winkler for Dissolved Oxygen by the Iodine Difference Method (OHLE)

Reagents

1. Iodine Stock Solution

Dissolve 5 gm iodine plus 100 KI in 114 ml water. One ml of the solution contains about 44 mg I_2 and 0.88 gm KI.

2. Iodine Reagent

Dissolve 35 gm NaCl in 100 ml distilled water; add 10 ml of the above Iodine Stock Solution. One ml contains about 4 mg I_2 and 80 mg KI.

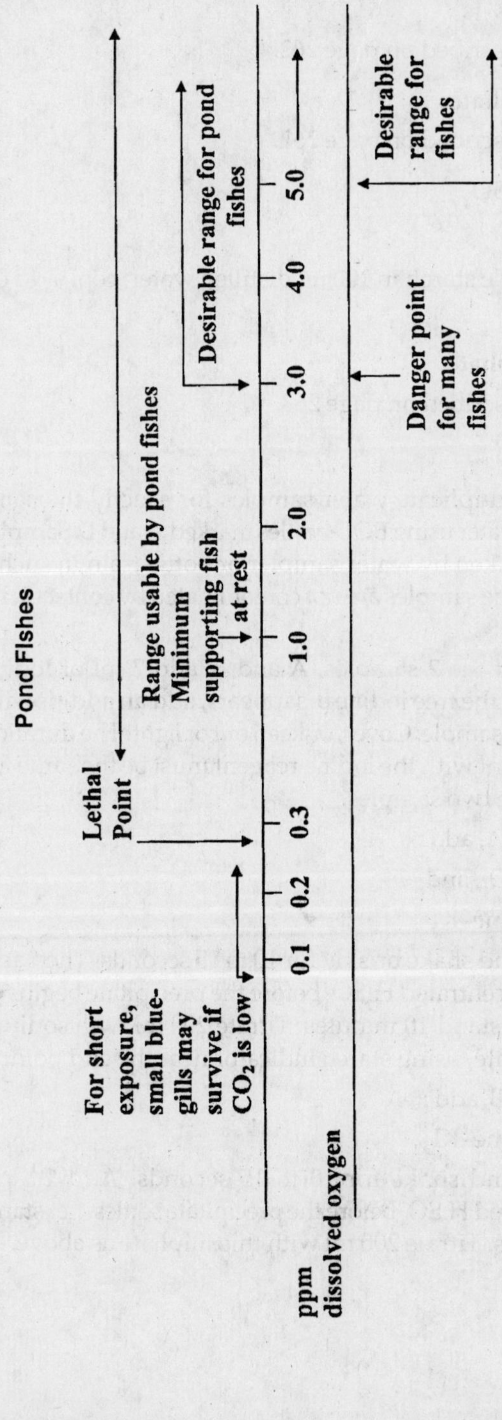

Figure 10.1: Oxygen Content of Water and its Relation to Fishes

3. Alkaline–KI

Prepared as described on page 262.

4. Manganous Sulphate

Prepared as described on page 262.

5. Concentrated H_2SO_4

6. Starch Solution

Boil 1 gm soluble starch in 100 ml distilled water; cool, add 0.5 per cent formalin.

7. Sodium Thiosulphate

Prepared as described on page 263.

Procedure

1. Collect 2 duplicate water samples for exactly the same depth of water using BOD bottles marked A and B. Samples must be taken by water sampler or bottle train in such a way that the samples are not contaminated by contact with air.
2. To each of the 2 samples, A and B, add 2 ml of Iodine reagent. If the free iodine disappears, add an additional 2 ml to each sample. Cover, or keep out of light. The duration of treatment with the iodine reagent must be the same for each of the two samples.
3. To bottle–A, add :

   1 ml $MnSO_4$, and

   3 ml alkaline–KI

   Stopper and shake briskly for 10 to 15 seconds. Then add 1.5 ml concentrated H_2SO_4 before the precipitate begins to settle. Let stand 10 minutes. Titrate 200 ml with sodium thiosulphate, using starch indicator near the end point.
4. To bottle–B, add :

   3 ml alkaline–KI

   Stopper and shake for 10 to 15 seconds. Add 1.5 ml concentrated H_2SO_4 before the precipitate settles. Let stand 10 minutes. Titrate 200 ml with thiosulphate as above.

5. The difference in titre of the 2 samples (A and B) is equivalent to the dissolved oxygen in the water sample plus the added in the reagents. The iodine equivalent to dissolved oxygen is liberated in bottle–A only. The titre for the second bottle (B) corrects for interfering substances liberating and /or absorbing iodine if the reaction time for the two is the same.
6. For most accurate determination of O_2, the proportionate volumes of reagents must be subtracted from the volume titrated and the dissolved oxygen in the reagents must be subtracted from the results of the titration. Correction for volume of reagent can be obtained by titrating 210 ml instead of 200 ml.

Portable Polarographic Oxygen Meter

Principles

The system consists of a gold electrode and a silver-silver oxide reference electrode. When a suitable polarizing voltage is applies across the cell, molecular oxygen is reduced to form OH^- at the gold electrode.

$$4\,e^- + O_2 + 2\,HOH \rightarrow 4\,OH^-$$

A current then flows to the silver electrode and the OH ions react with the silver to form Ag_2O

$$2\,Ag + OH^- \rightarrow Ag_2O + HOH + 2\,e^-$$

The Ag_2O appears as black or brown coating on the silver, and act to form a half-cell which completes the circuit with the gold electrode. The amount of current flowing between the two electrodes depends upon the amount of O_2 reaching the gold elɔctrode.

Since organic materials and metals accumulate upon a bare gold electrode and reduce its sensitivity. It is necessary to enclose the electrodes (bathed in 0.5 N KCl) within a teflon membrane. Molecular oxygen can diffuse through the membrane from the water, but ions cannot, membranes of a thickness from 0.5 to 1 ml are used. The current flow will be approximately 1/1000 of which would flow between the bare electrodes if no membrane were used. Since the oxygen must pass through the membrane and the O_2 inside the membrane is effectively zero, the apparatus measures the oxygen tension.

The Winkler analysis for dissolved oxygen (DO) determines oxygen in ppm, whereas, the oxygen meter measures oxygen tension.

Parts per million DO is the weight in milligram of dissolved oxygen per liter of water. Oxygen tension is the pressure of oxygen above a water surface required to maintain a certain weight of oxygen gas in solution when the two are at equilibrium – that is when the same numbers of molecular oxygen are leaving the water as entering from the air above per unit of time. This water is said to be "saturated" with oxygen.

Temperature, however, has a important effect upon the ppm dissolved oxygen in water at saturation. The higher the water temperature, the higher the partial pressure required to maintain, for example, 9 ppm dissolved oxygen. In pure water at 20°C and 760 air pressure water is "saturated" at 9.2 ppm oxygen. However, if air pressure remains the same and water temperature rises to 25° C, the water with 9.2 ppm DO is "supersaturated" and will lose oxygen to the air until 8.4 ppm is reached, where the water is "saturated" at the new temperature. However, since the air pressure did not change, and since oxygen is 20.92 per cent of air, the oxygen tension was 760 × 0.2092, or 160 mm in both cases. The oxygen meter consequently gives the same reading in saturated water at both temperatures.

In addition, the salt content of water has an effect similar to that of temperature. The higher the salt content, the greater the oxygen tension required to maintain a certain weight of oxygen in solution. At 760 mm pressure and 25° C pure water holds 9.2 ppm oxygen at saturation, while seawater (3.5 per cent salt) holds only 7.4 ppm. The oxygen meter will read approximately the same figure for both because the oxygen tension is the same.

It is thus evident that the oxygen meter must be calibrated for temperature and salinity to convert its readings into ppm.

Operation

These directions are written for use with a YSI model 51 oxygen meter but can be used with appropriate modifications with other machines. For best results read the instruction manual for the instrument to known how to use it properly.

Preparation of Probe

1. Remove guard, old membrane, and 'O' ring and rinse electrodes with distilled water or KCI solution.

2. Fill hole in probe with KCl and form a large drop over gold electrode.
3. Stretch membrane over electrodes, avoiding bubbles, and fasten down with 'O' ring. While handling, keep membrane clean and free of fingerprints.
4. Trim excess membrane and replace guard. Probe is ready for operation.

Procedure

1. Turn selector switch to red line and adjust needle to the line with slotted control knob. If you cannot adjust needle to the line the batteries need replacing (see instrument instruction manual).
2. Turn selector to zero and adjust needle to zero.
3. Lower probe into water, turn selector to temperature and read temperature. Allow 30 second to 1 minute for equilibration.
4. Turn selector to read, gently move probe or otherwise cause water to flow across membrane, and read O_2. Low readings will result if there is no flow across the membrane.

Calibration

The meter may be calibrated against a Winkler determination of dissolved oxygen in the water or for fresh waters it may be calibrated in air. For greatest accuracy, calibration should be done by the Winkler analysis in waters of the same salinity and at the same temperature (+3°C) as the measurements to be made. Where this impossible, corrections can be made but the accuracy is lowered.

To Calibrate in Air

1. Place probe in damp calibration chamber and place in water to be measured.
2. Allowed temperature to equilibrate (3 to 5 minutes).
3. Read temperature and determine calibration value (saturation) for the temperature and atmospheric pressure from calibration table.
4. Turn selection to read and adjust meter to calibration value.

To Calibrate Against Winkler Determination

1. Determine oxygen by Winkler Method.
2. Insert probe, turn selector to read and adjust meter to value determined in step 1.

Notes

Accuracy

The accuracy is greatest (±0.3 ppm) when calibration and measurements are made at the same temperature. When the temperature difference reaches 10°C, the error may be as great as 2 ppm. When the O_2 is below 1 ppm, the meter readings are only approximate. It is also inaccurate in supersaturated waters.

Table 10.1: Calibration Tables Solubility of Oxygen in Water (saturated with air) in ppm at Various Temperature and Pressure

P mm	*775*	*760*	*750*	*725*	*700*	*675*	*650*	*625*
P inches	*30.51*	*29. 92*	*29.53*	*28.54*	*27.56*	*26.57*	*25.59*	*24.61*
°C	ppm							
0	14.9	14.6	14.4	13.9	13.5	12.9	12.5	12.0
1	14.5	14.2	14.1	13.6	13.1	12.6	12.2	11.7
2	14.1	13.9	13.7	13.2	12.9	12.3	11.8	11.4
3	13.8	13.5	13.3	12.9	12.4	12.0	11.5	11.1
4	13.4	13.2	13.0	12.5	12.1	11.7	11.2	10.8
5	13.1	12.8	12.6	12.2	11.8	11.4	10.9	10.5
6	12.7	12.5	12.3	11.9	11.5	11.1	10.7	10.3
7	12.4	12.2	12.0	11.6	11.2	10.8	10.4	10.0
8	12.1	11.9	11.7	11.3	10.9	10.5	10.1	9.8
9	11.8	11.6	11.5	11.1	10.7	10.3	9.9	9.5
10	11.6	11.3	11.2	10.8	10.4	10.1	9.7	9.3
11	11.3	11.1	10.9	10.6	10.2	9.8	9.5	9.1
12	11.1	10.8	10.7	10.3	10.0	9.6	9.2	8.9
13	10.8	10.6	10.5	10.1	9.8	9.4	9.1	8.7
14	10.6	10.4	10.2	9.9	9.5	9.2	8.9	8.5
15	10.4	10.2	10.0	9.7	9.3	9.0	8.7	8.3

Contd...

Table 10.1–Contd...

P mm	*775*	*760*	*750*	*725*	*700*	*675*	*650*	*625*
P inches	*30.51*	*29. 92*	*29.53*	*28.54*	*27.56*	*26.57*	*25.59*	*24.61*
°C	ppm							
16	10.1	9.9	9.8	9.5	9.1	8.8	8.5	8.1
17	9.9	9.7	9.6	9.3	9.0	8.6	8.3	8.0
18	9.7	9.5	9.4	9.1	8.8	8.4	8.1	7.8
19	9.5	9.3	9.2	8.9	8.6	8.3	8.0	7.6
20	9.3	9.2	9.1	8.7	8.4	8.1	7.8	7.5
21	9.2	9.0	8.9	8.6	8.3	8.0	7.7	7.4
22	9.0	8.8	8.7	8.4	8.1	7.8	7.5	7.2
23	8.8	8.7	8.5	8.2	8.0	7.7	7.4	7.1
24	8.7	8.5	8.4	8.1	7.8	7.5	7.2	7.0
25	8.5	8.4	8.3	8.0	7.7	7.4	7.1	6.8
26	8.4	8.2	8.1	7.8	7.6	7.3	7.0	6.7
27	8.2	8.1	8.0	7.7	7.4	7.1	6.9	6.6
28	8.1	7.9	7.8	7.6	7.3	7.0	6.7	6.5
29	7.9	7.8	7.7	7.4	7.2	6.9	.6.6	6.4
30	7.8	7.7	7.6	7.3	7.0	6.8	6.5	6.2
31	7.7	7.5	7.4	7.2	6.0	6.7	6.4	6.1
32	7.6	7.4	7.3	7.0	6.8	6.6	6.3	6.0
33	7.4	7.3	7.2	6.9	6.7	6.4	6.2	5.9
34	7.3	7.2	7.1	6.8	6.6	6.3	6.1	5.8
35	7.2	7.1	7.0	6.7	6.5	6.2	6.0	5.7
36	7.1	7.0	6.9	6.6	6.4	6.1	5.9	5.6
37	7.0	6.8	6.7	6.5	6.3	6.0	5.8	5.6
38	6.9	6.7	6.6	6.4	6.2	5.9	5.7	5.5
39	6.8	6.6	6.5	6.3	6.1	5.8	5.6	5.4
40	6.7	6.5	6.4	6.2	6.0	5.7	5.5	5.3

Dissolved Salts

Since the meter measures oxygen tension and not absolute quantity and since the solubility of gases decreases with increased dissolved salts, the meter will give high readings as ppm DO in saline waters if calibrated in air without correcting for salt content. A change from O (distilled water) to 3.5 per cent salts (sea water) decreases O_2 solubility by about 20 per cent so the meter will read about 20 per cent high. The dissolved salts in most fresh waters do not cause a significant change from distilled water. Calibration against a Winkler determination eliminates this source of error in saline waters.

Acidity and Alkalinity of Water

Determination of Free CO_2 and HCO_3^-, CO_3^{--} and OH^- Alkalinities

The titration method is described here for these determinations. A nomographic method of determination may be found in APHA Standard Methods.

The approximate relationships of CO_2 and HCO_3^-, CO_3^{--} and OH^- alkalinities are shown on the accompanying chart.

Separation of the various forms of CO_2 is based on the assumption that the end point of phenolphthalein (approximately 8.3), gives a sufficiently satisfactory division between carbonates and bicarbonates for most biological purposes. In more alkaline waters from pH 8.3 to 10.5, there may be a mixture of HCO_3^- and CO_3^{--} but the bicarbonate decreases sharply as the pH value approaches 10.4. At higher pH values relatively insignificant amounts of bicarbonate are present. In waters more acid the 8.3, practically no carbonate is present, but only varying mixtures of HCO_3^- and CO_2.

In waters that are exposed to the air and are more acid than pH 5.5, acidity is assumed to be due to stronger acids that CO_2. When at equilibrium with air at atmospheric pressure of 760 mm, water contains only approximately 0.4 ppm free CO_2 and can not make the water more acid than 5.5. However, water from deep wells, from springs or the deeper waters of lakes or ponds where CO_2 gas is released by decomposition, are exposed to much high partial pressures of the gas and may contain up to 200 or more ppm of free CO_2, with correspondingly lower pH values. Upon exposure to air, the free CO_2 is rapidly lost until equilibrium with the partial pressure of CO_2 in the atmosphere is reached. CO_2 can be rapidly removed from water by heating the water to boiling.

Titration with either Na_2CO_3 or NaOH to the phenolphthalein end point allows estimation of free CO_2 in water.

Principles

1. Water Sample Acid to Phenolphthalein and Alkaline to Methyl Orange:

Contains free CO_2 and bicarbonates.

a. Determine free CO_2 by titration with Na_2CO_3 with phenolphthalein indicator.

$$CO_2 + HOH + Na_2CO_3 \rightarrow 2\,NaHCO_3$$

The equivalent weight of Na_2CO_3 in this reaction is the molecular weight divided by two.

b. Determine the HCO_3^- in a fresh sample by titration with HCl, with methyl orange indicator.

$$HCl + NaHCO_3 \rightarrow NaCl + HOH + CO_2$$

2. Water Sample Alkaline to Phenolphthalein (Pink Colour)

Contains CO_3^{--}, HCO_3^- or possible OH^-

a. Determine phenolphthalein alkalinity by titration with standard HCl to phenolphthalein and point (pH = 8.3).

$$Na_2CO_3 + HCl \rightarrow NaHCO_3 + NaCl$$

The equivalent weight of Na_2CO_3 is the molecular weight, since 1 molecule combines with 1 equivalent of HCl.

b. Using the colourless solution from the above titration, determination the bicarbonate alkalinity by titration with standard acid to the methyl orange or the methyl orange-xylene cyanole end point (pH = 4.5).

$$NaHCO_3 + HCl \rightarrow NaCl + HOH + CO_2.$$

This bicarbonate is a mixture of the bicarbonate originally present and that formed in the preceding titration from the carbonate, if present. The latter will require the same volume of acid as was used in the original titration under Section a. If this amount is subtracted from the volume used in the second titration, the difference is that required by the bicarbonate originally present.

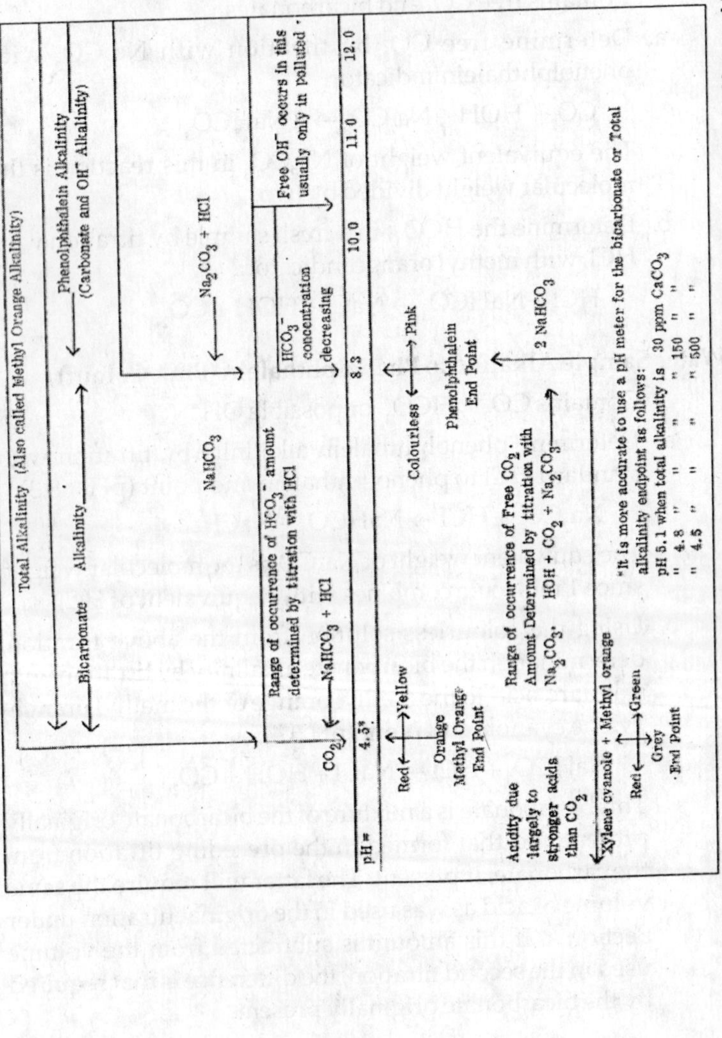

Figure 10.2: Relationship of Determination of CO_2, HCO_3^-, CO_3^{--} and OH^- in Natural Water

Interpretation

a. Total alkalinity is due not to carbonates but to OH^- when a sample has phenolphthalein alkalinity but no bicarbonate alkalinity, because the titration of carbonate to the phenolphthalein end point would have produced bicarbonate.

b. Total alkalinity is due to only HCO_3^- if there is bicarbonate alkalinity but no phenolphthalein alkalinity.

c. Total alkalinity is due to only CO_3^{--} if the bicarbonate alkalinity equals the phenolphthalein alkalinity.

d. Total alkalinity is due to OH^- and CO_3^{--} when phenolphthalein alkalinity is greater than the bicarbonate alkalinity.

Reagents

1. Standard 0.0454 N Na_2CO_3 Solution

The molecular weight of sodium carbonate is 106, and since it contains 2 equivalents, a 1.0 N solution contains $\frac{106.0}{2} = 53.0$ gm per liter or 26.5 gm per 500 ml.

Preparation of Stock Standard Solution of Na_2CO_3

a. Heat a clean 100 ml evaporating dish to 200° C in an oven for 30 minutes. (or heat at 120° C for 10 hours). Cool in a desiccator. Remove and weigh accurately.

b. Add approximately 30 gm of C.P. reagent-grade Na_2CO_3.

c. Heat to 200° C in an oven for 30 minutes. Stir and reheat for 30 minutes. Cool in a desiccator. Remove, weight accurately.

d. Dissolve in freshly boiled and cooled distilled water; transfer to a 500 ml volumetric flask and dilute to the mark with the boiled and cooled distilled water.

e. Calculate the actual normality of this stock solution as follows:

$$N = \frac{\text{gm of } Na_2CO_3}{26.5}$$

For example, if the weight in Section 3 was 27.162 gm Na_2CO_3 the stock solution was :

$$N = \frac{27.162}{26.5} = 1.025\,N$$

Preparation of 0.0454 N Na_2CO_3 from the Stock Solution

f. The ml of stock standard solution required to make 1 liters of 0.0454 N Na_2CO_3 is calculated as follows :

$$\frac{0.0454\ N}{\text{Normally of stock solution}} \times 1000$$

For example, if the stock solution was 1.025 N Na_2CO_3,

then $\frac{0.0454}{1.025} \times 1000 = 44.3$ ml of 1.025 N Na_2CO_3 required to make liter of 0.0454 N Na_2CO_3.

g. Measure accurately the required number of ml of the stock solution from a burette in a 1 liter volumetric flask*. Dilute to the mark with freshly boiled and cooled distilled water. Stopper tightly with a rubber stopper.

(* The indicator may be included in the 0.0454 N Na_2CO_3 by adding 16 ml phenolphthalein indicator before dilution to 1 liter).

Alternate

0.0227 N NaOH may be used in place of 0.0454 N Na_2CO_3 where a standard NaOH is available. Titration and calculations are the same with both.

2. Standard 0.0167 N HCl

This standard is most accurately prepared from a constant boiling mixture as follows :

Stock Solution of Standard HCl

a. Place 100 ml pure concentrated HCl and 400 ml distilled water in a beaker.

b. Boil under hood until half the solution has evaporated.

c. Record the barometric pressure during boiling.

d. The remaining solution from (b) will have a constant composition as indicated in the following table :

Barometric Pressure	*Percent HCl in Solution*	*Grams Solution* Required for 1 Liter of 1.0 N HCl*
770	20.197	180.407
760	20.221	180.193
750	20.245	179.979
740	20.269	179.766
730	20.293	179.555

* Corrected for weight in air.

Values for intermediate barometric pressure are determined by interpolation.

e. Weigh accurately and rapidly approximately 190 ml of the acid solution and place in a glass–stoppered volumetric flask. Dilute to 1 liter with recently boiled and cooled distilled water.

f. Determine the normality of the above stock standard solution as follows :

$$N = \frac{\text{Actual weight of HCl solution}}{\text{gm solution required for 1N HCl (from Table, Step - d)}}$$

For example, if the barometric pressure during boiling was 740 (Step-c) then 179.766 gm of the solution was required to make 1 liter of 1.0 N HCl (Step-d). If the actual weight of the acid solution was 195.120 (step-e), then the normality of the stock solution was:

$$\frac{195.120}{179.766} = 1.085\text{N}$$

Preparation of 0.0167 N HCl from the Standard Stock Solution

g. The ml of standard stock solution required to make 1 liter of 0.0167 N HCl is calculated as follows :

$$\frac{0.0167}{\text{Normality of stock solution}} \times 1000$$

For exainple, if the stock solution is 1.085 N HCl, then,

$$\frac{0.0167}{1.085} \times 1000 = 15.4 \text{ ml}$$ of 1.085 N HCl, is required to make 1 liter 0.0167 N HCl.

h. Measure accurately from a burette the required number of ml of the standard stock solution into 1-litter volumetric flask with a ground glass stopper. Dilute to the mark with freshly boiled and cooled distilled water.

3. Indicator : Make either a or b

a. Methyl Orange – Xylene Cyanole Indicator

Dissolve 0.2 gm methyl orange plus 0.2 gm xylene cyanole in 100 ml distilled water. Its end point is a grey colour at pH 4.5. In more alkaline waters it is green and in more acid waters, it is red. The end point is easy to detect.

b. Methyl Orange Indicator

Dissolve approximately 0.2 gm methyl orange in 100 ml water. This indicator has its transition point at pH 4.5 and at this point has an orange colour. In a more acid range it is red and in a more alkaline range, it is yellow. The end point is difficult to detect.

4. Phenolphthalein Indicator

Dissolve approximately 0.05 gm phenolphthalein in 25 ml of 50 per cent alcohol. This indicator gives a pale pink colour at pH 8.3, its transition point. In more acid solutions it is colourless and in more alkaline solution it is red.

Procedures for Free Carbon Dioxide

Pretest 2 small water samples of approximately 20 ml by adding several drops of methyl orange-xylene cyanole (or methyl orange) to one and several drops phenolphthalein to the other. If the sample is acid to methyl orange-xylene cyanole, acidity is due to acids stronger than CO_2. If the samples are alkaline to methyl orange and acid (colourless) to phenolphthalein, the acidity is assumed to be due to carbon dioxide.

1. In the latter case, carefully pipette 200 ml of the water sample into an Erlenmeyer flask.

2. Add 10 drops phenolphthalein (The solution should remain colourless). If the standard Na_2CO_3 contains the indicator, addition of phenolphthalein to the sample is unnecessary.
3. Titrate to a faint pink (approximately pH 8.3) with 0.0454 N sodium carbonate.
4. The number of ml of 0.0454 N sodium carbonate x 5 = ppm free CO_2.
5. To differentiate between acidity due to free CO_2 and that due to strong acids (usually mineral acids), proceed as follows :
   a. Pipette duplicate water samples of 200 ml in each of 2 flasks.
   b. Bring one sample to boiling and cool.
   c. Add 10 drops phenolphthalein to each sample.
   d. Titrate each sample to a faint pink (pH 8.3) with 0.0454 N Na_2CO_3. Difference in the titres unboiled sample-boiled sample is acidity due to CO_2. Calculate as in 4 above. The titre of the unboiled sample represent total acidity, while that of the boiled sample usually is due largely to mineral acids. Sulphuric acid is the most common mineral acid found in natural waters.

Procedures for Alkalinity, CO_3^{--}, HCO_3^-, and OH^-

1. Determination of Phenolphthalein Alkalinity

a. To 100 ml of the sample water, add a few drops of phenolphthalein.

b. If, the solution remains colourless, no phenolphthalein alkalinity is present.

   Continue with Section 2-a.

c. If, a pink colour develops upon the addition of phenolphthalein titrate with 0.0167 N HCl, adding a drop every 2 to 3 seconds until the pink colour disappears. Record the burette reading as phenolphthalein alkalinity (P).

d. Save the sample for titration under Section 2-b.

2. Determination of bicarbonate alkalinity :

a. To the water sample from 1-b above, add several drops methyl orange or methyl orange-xylene cyanole and titrate with 0.0167 N HCl to the end point (orange colour, or grey colour respectively). Record the burette reading as methyl orange titre (M).

b. To the water sample from 1-c above, add several drops of methyl organe-xylene cyanole or methyl orange indicator and titrate with 0.0167 N HCl to the end point. Record burette reading as methyl orange titre (M). If, the titre is large, it is best to titrate to the end point, then boil to expel CO_2, and retitrate to the end point. The presence of large amounts of CO_2 may give a premature end point if this is not done.

Calculations

A. If; M is larger than P, subtract M-P = X, then :

1. Multiply P ml by 10 to obtain the concentration of CO_3^{--} ions as ppm in the water sample.
2. Multiply X ml by 10.1 to obtain the concentration of HCO_3^- ions as ppm in the same sample.

B. If; M=P, then there is no bicarbonate (HCO_3^-) alkalinity, but only CO_3^{--} alkalinity. Multiply P ml by 10 to obtain the concentration of CO_3^{--} as ppm in the water sample.

C. If; P is larger than M, subtract P-M = Y then :

1. Multiply M by 10 to obtain the concentration of CO_3^{--} ions as ppm in the water sample.
2. Multiply Y by 2.84 to obtain OH^- as ppm, or multiply by 5.0 to express OH^- alkalinity as the equivalent in ppm CO_3^{--}.

   Total alkalinity expressed as CO_3^{--} equivalent is equal to 5 (M + P).

   Total alkalinity expressed as $CaCO_3$ equivalent is equal to 8.35 (M + P).

pH

There is only one satisfactory method of determining pH of pond waters and that is by the use of the electrometric glass electrode. Laboratory and portable models are available.

Universal indicator dyes are available which give different colours over the entire range of pH. The colour produced is then compared to a colour chart. Errors in determining pH by the method in natural waters vary from 1.0 to 3.0.

Colourimetric determinations using pH indicators, with solutions of known pH for comparison are more accurate, but certain waters may give colours varying 1.0 from the true pH value.

Overlapping indicators, as shown on the following sheet (Figure 10.4), may be used for approximations of pH, but are subject to errors of the type mentioned above.

Indicator papers are inaccurate, especially in poorly buffered waters. They are useful only to obtain approximate pH values.

For purposes of water classification, the pH of waters should be taken preferably just before daybreak, or before active photosynthesis has materially reduced the available free and/or half-bound carbon dioxide.

Ammonia Nitrogen

Soluble ammonia nitrogen forms a yellow to orange colour with Nessler's reagent, the depth of colour being proportional to the amount of NH_4^+ present.

Reagents

1. Standard Ammonium Chloride Solution Containing 5 ppm N

Since the amount of NH_4Cl required to make a liter of solution containing 5 ppm N (or 0.005 gm N per liter) is small,

$$\begin{matrix} NH_4Cl \rightarrow N \\ 53.3 \qquad 14 \end{matrix}$$

$$0.005 \text{ gm} \times \frac{53.3}{14} = .019105 \text{ gm } NH_4Cl,$$

it is better to prepare first a more concentrated stock solution.

a. Weigh accurately 0.019105 gm of pure NH_4Cl. Dilute to 1 liter with ammonia-free distilled water.

b. Measure with a volumetric pipette 10 ml of the above stock solution into a glass-stoppered volumetric flask and dilute to 1 liter with ammonia-free distilled water. This standard solution contains 5 ppm N in the form of NH_4^+.

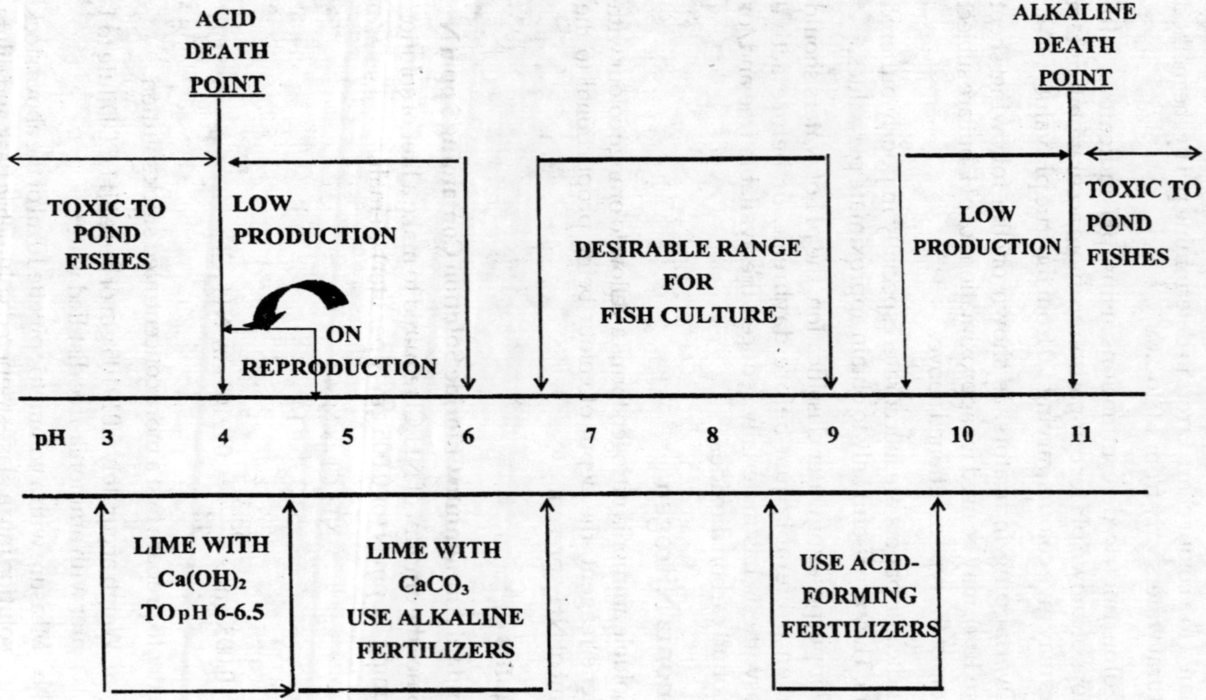

Figure 10.3: Relationship of pH of Pond Waters to their Suitability for Fish Culture

Indicator	Colour change	pH
Thymol Blue	Red – Orange Yellow	1.8 – 2.4
Methyl Orange	Red – Orange	3.1 – 4.4
Brom Phenol Blue	Yellow – Purple	3.4 – 3.8
Methyl Red	Red – Yellow	4.8 – 6.0
Brom Cresol Purple	Yellow – Purple	5.4 – 6.4
Brom Thymol Blue	Yellow – Green – Blue	5.4 – 6.4 – 7.0
Phenol Red	Yellow – Pink	6.8 – 7.4
Cresol Red	Yellow – Purple	7.0 – 8.2
Phenolphthalein	Colorless – Pink	8.0 – 9.0
Thymol Blue	Yellow-green – Blue	8.4 – 9.0
Thymolphthalein	Colorless – Blue	9.0 – 10.5

Figure 10.4: Overlapping Indicators, their pH and Colour Change in Pond water

2. Ammonia-Free Distilled Water

Test distilled water with Nessler's reagent as described below. If no ammonia N can be detected, it may be used directly. If traces of ammonia N are present, acidify with 5 ml concentrated H_2SO_4 per liter. Distill off three-fourths the water into a glass-stoppered bottle and label. The ammonia will remain in the acid water.

3. Nessler's Reagent

This reagent may be purchased ready-made or may be made as follows :

a. Dissolve 100 gm HgI_2 and 70 gm KI in 200 ml distilled water.

b. Dissolve 160 gm NaOH in 500 ml distilled water, allow to cool.

c. Add HgI_2 plus KI solution slowly, with stirring, to NaOH solution.

d. Dilute to 1 liter and store in pyrex out of direct sunlight. It should be stable for up to one year.

Determination

(For waters free of hydrogen sulphide and other interfering substances.)

1. Filter water sample to remove particulate matter.
2. Pour 50 ml of filtered water sample into 50 ml Nessler tubes.
3. Freshly prepare in 50 ml Nessler tubes 50 ml of standard solutions containing 0.1, 0.25, 0.5 and 1.0 ppm N as NH_4^+ by diluting the 5 ppm standard prepared under 1-b with ammonia-free distilled water as follows :

For Standards with Following ppm N	*Dilute Following Amounts of Standard NH_4Cl Solution (Containing 5 ppm N) to 50 ml*
0.10	1.0 ml
0.25	2.5 ml
0.50	5.0 ml
0.75	7.5 ml
1.0	10.0 ml

4. Add to each sample and to each of the above standard 2 ml Nessler's reagent. Invert to mix and let stand 10 minutes for full colour to develop.
5. Compare the water sample with the standards. Estimate the ammonia as ppm N.

   The comparison should be made by placing a light behind the rack containing the Nessler tube and comparing only the colour observed by looking from the side of the tube directly across the meniscus at the top of the liquid column. This minimizes errors due to precipitates and off-colours in the sample and sometimes will be found to give results approximating those obtained by distillation of the ammonia and subsequent nesslerization.
6. If the sample contains more than 1.0 ppm N, a second sample should be taken, diluted with 4 parts or more of ammonia-free water, and subsequently nesslerized as in Section 4.

Interfering Substances

Hydrogen sulphide, sulphides, SO_2, amines, proteins or aldehydes, may cause greenish off-colours. Calcium, magnesium and iron may cause precipitates or turbidities. Interference from these sources can sometimes be overcome by the following pre-treatment of the water. (However, it is preferable to use distillation).

Regents

1. *$ZnSO_4$*: Dissolve 10 gm $ZnSO_4 . 7 H_2O$ in ammonia-free water and dilute to 100 ml.
2. *NaOH*: Dissolve 25 gm NaOH in ammonia-free water and dilute to 100 ml.
3. *Rochelle salt solution*: Dissolve 50 gm of potassium-sodium titrate ($KNa C_4H_4O_6 . 4 H_2O$) in 100 ml and boil until free of ammonia. Dilute to 100 ml.
4. *Lead acetate*: Dissolve 2 gm in 100 ml distilled water.

Procedure

1. Remove residual free chlorine with a few drops of 0.025 N sodium thiosulphate (oxygen reagent).

2. Add several drops lead acetate solution and 1 ml $ZnSO_4$ solution to 100 ml of the water sample. Mix.
3. Add. 0.5 ml NaOH to a pH of 10.5. Mix. The precipitate should fall, clearing the water and removing proteins, amino acids, sulphides and SO_2.
4. Remove 50 ml of the supernatant liquid to a Nessler tube, add 5 to 6 drops of Rochelle salt solution to remove Ca and Mg.
5. Nesslerize the above solution and compare with standards.

Distillation

Where interfering substance are present, the most accurate method is to remove to ammonia by distillation. Since this is time-consuming, it is not ordinarily done unless very accurate results are needed, or the methods given above do not work.

Reagents

1. *Phosphate buffer (pH 7.4)*: Dissolve 14.3 gm KH_2PO_4 and 90.1 gm K_2HPO_4. 3 H_2O is distilled water and dilute to 1 liter.
2. *Dilute HCl*: Dilute 8.5 ml conc. HCl to 1 liter (approximately 0.1 N).
3. *10 per cent NaOH*: Dissolve 100 gm NaOH in distilled water and dilute to 1 liter.

Procedure

Use the Kjeldahl Distillation setup as shown on the following.

1. Place 200 ml of the water sample in a distillation flask.
2. Neutralize to litmus paper with a few drops of 10 per cent NaOH. Add a few boiling beads or chips and 5 ml phosphate buffer to give a pH of 7.4. In very hard water (250 ppm Ca or more) more buffer may be needed : add up to 20 ml buffer and adjust pH to 7.4 with acid or base.
3. Distill into collecting flask by bubbling through 10 ml of the dilute HCl until you have collected about 80 ml of the distillate.

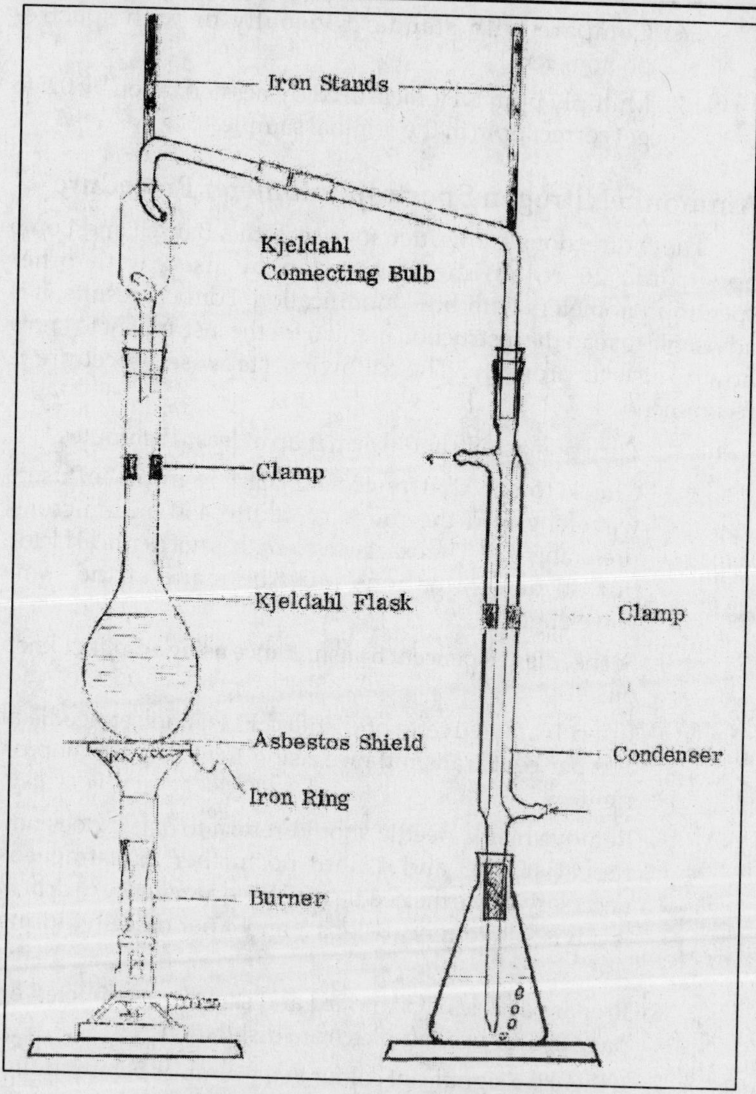

Figure 10.5: Kjeldahl Distillation Setup

4. Dilute to 100 ml, remove 560 ml and nesslerize.
5. Save residue in distillation flask for organic nitrogen determination.

6. Compare with standard visually or with spectrophotometer.
7. Multiply ppm NH_3 measured by nesslerization by 0.5 to get correct ppm in the original sample.

Ammonia Nitrogen Spectrophotometer Procedure

These directions are written for use with a Bausch and Lomb Spectronic 20 colourimeter but can be used with other spectrophotometers with little modification. For best results, it is advisable to read the instruction manual for the instrument to know how to use to properly. The following stepwise procedure is recommended:

1. Turn of machine and warm it up at least 15 minutes.
2. Check to see that photocell 5581 is in place, use a wavelength in the range of 400 to 450 millimicrons, preferably 425. The exact wavelength is not a critical factor, but all determinations must be made at the same wavelength.
3. Set needle at 0 per cent transmittance using amplifier knob on left.
4. Insert blank (cuvette of distilled H_2O) and set needle at 100 per cent transmittance using light control knob on right.
5. Remove blank, needle should return to 0. If it does not, repeat steps 3 and 4 until no further adjustment is necessary. Since the needle may have a tendency to drift, it is advisable to repeat steps 3 and 4 periodically during use.
6. Prepare a series of standard and samples as directed on page 286, or use samples from distillation.
7. If there is a significant colour in the samples before adding Nessler's reagent, place a portion of the sample in cuvette and read per cent transmittance. This is the sample blank.
8. Since colour development changes with time, the development time between adding Nessler's reagent and reading the colour should be constant. Ten minutes is recommended. Add the Nessler's reagent to the first sample and mix. Record the time. After a suitable interval, add

Nessler's reagent to the second sample. Repeat with each sample at the same interval. The interval is determined by the speed at which the operator can read the spectrophotometer in step 9 (using a single cuvette an operator can read a simple every 30 seconds with a little practice).

9. Ten minutes after adding Nessler's reagent to the first sample, transfer it to a cuvette and read the per cent transmittance. Repeat with each sample using the interval between samples used to step 8. In this way, the colour develops for ten minutes in each sample.
10. Convert percent transmittance to optical density using the table in the Appendix.
11. Plot optical density of the standard against ppm of the standards on regular graph paper. The calibration curve is the straight line that best fits the points. Record wavelength and development time used on graph for future use of calibration curve. Above about 1.2 to 1.5 ppm, the straight line relationship does not hold. More concentrated samples should be diluted to bring them below this range before nesslerization.
12. Subtract optical density of sample blank from that of sample. Read ppm of samples reagent blanks directly from calibration curve. Subtract ppm of reagent black from that of sample after adjusting sample ppm for whatever dilution of the sample was necessary.

Organic Nitrogen

Reagents

All the reagents listed for the distillation and nesslerization of ammonia (page 283, 286 and 288) are required, plus the following;

1. *Digestion reagent a.*: Dissolve 134 gm K_2SO_4 is 650 ml H_2O + 200 ml conc. H_2SO_4. b. : Dissolve 2 gm H_gO in 25 ml 6 N H_2SO_4. Add 'b' to '2' with stirring, dilute to 1 liter. Store above 14° C.
2. *Phenolphthalein Indicator*: Dissolve 0.5 phenolphthalein in 100 ml distilled water. Add dilute NaOH until faint pink colour appears.

3. *NaOH – Na thiosulphate reagent*: Dissolve 500 gm NaOH and 25 gm $Na_2S_2O_3$. 5 H_2O in 1 liter distilled water.

Procedure

1. Follow procedures for distillation of NH_3 and save residue in Kjeldahl flask.
2. To cooled residue acid 50 ml digestion reagent and mix.
3. Heat under hood until white SO_3 fumes are given off and sample clears.
4. Digest for additional 30 minutes and allow to cool.
5. Dilute with 200 ml H_2O, and 0.5 ml phenolphthalein indicator and mix.
6. Add 50 ml hydroxide-thiosulphate reagent, pouring down side of flask.
7. Connect to distillation apparatus and stir to mix the 2 layers. Mixture will be black with a red tint. Add more hydroxide-thiosulphate reagent if red colour does not appear.
8. Distill, collect about 130 ml in 20 ml 0.1 N HCl (or H_2SO_4) and dilute to 200 ml.
9. Simultaneously subject 200 ml nitrogen-free water to same procedure to be used as reagent-blank.
10. Nesslerize and evaluate concentration by spectrophotometer as per ammonia determination.

Nitrite Nitrogen

Since nitrites may be oxidized readily to nitrates, samples of water should be analyzed promptly. The sulphanilic acid methods is used.

Reagents

1. *Sulphanilic acid solution**: Dilute 30 ml glacial acetic acid to 100 ml with distilled water. Add 1 gm sulphanilic acid.
2. *Alpha-naphthylamine solution**: Dilute 30 ml glacial acetic acid to 100 ml with distilled water. Add 0.5 gm alpha-naphthylamine. The reagent is not stable.
3. *Stock $NaNO_2$ solution*: Dissolve 1.230 gm $NaNO_2$ in 500 ml distilled water.

4. *Standard $NaNO_2$ solution*: Using volumetric pipette dilute 50 ml of the stock $NaNO_2$ solution to 500 ml. Then exactly 5 ml of this dilute solution and dilute again to 500 ml. This solution contains 0.5 ppm N as NO_2^-.

   (* A dry reagent may be prepared by grinding 1 gm alpha-naphthylamine, 10 gm sulphanilic acid and 89 gm tartaric acid in a mortar. Use 0.3 gm in 50 ml water sample. This reagent is stable.)

Procedure

1. Filter 50 ml of the water sample into a Nessler tube.
2. Add 1 ml sulphanilic acid solution.
3. Add 1 ml alpha-naphthylamine solution and mix. This gives a pink to red colour within 10 minutes, if NO_2^- is present.
4. Compare with standard diluted from the standard $NaNO_2$ solution and similarly treated.

Number of ml Standard $NaNO_2$	*ppm N when Diluted to 50 ml*
1	0.01
2	0.02
3	0.03
4	0.04
5	0.05

If measured in a spectrophotometer, set the wavelength at 530.

Nitrate Nitrogen

The phenoldisulphonic acid method is used. Nitrates react to form the tripotassium salt of nitrophenoldisulphonic acid.

If the chloride content is greater than 30 ppm, the chlorides should first be precipitated by adding several drops of silver sulphate (5 per cent) and filtering.

Nitrites, if in excess of 0.2 ppm should first be determined and converted to nitrates by heating with 5 ml H_2O_2 to each 100 ml of water sample. The amount of nitrate (NO_3^-) formed by oxidation of NO_2^- should be subtracted from the final results.

Reagents :

1. *Phenoldisulphonic acid solution*: This acid may be purchased or prepared as follows: Dissolve 25 gm of pure white phenol in 150 ml of concentrated sulphuric acid, add 75 ml of fuming sulphuric acid (13-15 per cent SO_3), and heat at 100° C for 2 hours. Store in a brown bottle.
2. *Standard nitrate solution*: Dissolve 0.607 gm of pure sodium nitrate in 1 liter of nitrate-free distilled water. Evaporate 50 ml of this solution to dryness in a porcelain dish; when cool, treat with 2 ml of the phenoldisulphonic acid solution, rubbing with a glass rod to insure intimate contact, and dilute to 500 ml with distilled water. One ml is equivalent to 0.01 mg of nitrogen as nitrate. (This solution is permanent) Prepare standards for comparison by adding strong ammonium hydroxide (1-1) to measure volumes of the standard solution in 100 ml Nessler tubes. The intensity of the yellow colour increases with increasing NO_3^- concentration.
3. *Ammonium hydroxide*: Dilute 500 ml NH_4CH to 1 liter.
4. Prepare standard as follows :

Standard Nitrate	*Diluted to*	*Equal to ppm*
0.1 ml	100 ml	0.01
0.5 ml	100 ml	0.05
1.0 ml	100 ml	0.10
2.0 ml	100 ml	0.2
5.0 ml	100 ml	0.5
10.0 ml	100 ml	1.0

Determination

1. Filter 100 ml water sample and place in porcelain evaporating dish.
2. Place on steam bath and evaporate to dryness.
3. Cool, and 2 ml phenoldisulphonic acid and mix with a glass rod.
4. Dilute with approximately 10 ml water.

5. Add NH_4OH (1-1) until maximum colour develops. Transfer to 100 ml Nessler tube and dilute with distilled water to the mark.
6. Compare with standards prepared in Section 4 above visually or with spectrophotometer.

Nitrate Nitrogen Determination by Spectrophotometer

1. Prepare standards of known concentration as described in Reagents section 4. Concentrations of 0.01, 0.05, 0.10, 0.15 and 0.20 ppm are suggested.
2. Switch on spectrophotometer to warm up machine for at least 15 minutes. Using a wavelength within a range of 400 of 450 millimicrons, with 5581 photocell, set machine to zero optical density by using distilled water. Subsequently, take optical density readings for the known standards. From the values obtained compute conversion factor by dividing each concentration by corresponding optical density. The average of these quotients is the conversion factor (F). Also, plot optical density and concentration on regular graph paper.
3. To analyze unknown sample, follow steps 1-5 of determination on page 294 and 295.
4. Transfer treated sample obtained in step 5 to cuvette and take optical density reading after setting machine to zero optical density with distilled water. For blank, use sample untreated with NH_4OH.
5. Correct the sample reading by subtracting the blank reading and read to concentration directly from the graph or multiply the corrected optical density reading by the conversion factor (F) to give to concentration of nitrates in ppm of the sample.

Total Nitrogen

Total nitrogen concentration in the water includes that in both inorganic and organic compounds, but excludes elemental N gas.

Reagents

1. *Concentrated sulphuric acid–salicylic acid solution*: Measure 900 ml of conc. Sulphuric acid into a large clean beaker.

Add 72 gm salicylic acid, stir the salicylic acid powder carefully into the acid until completely dissolved.

2. *Sodium thiosulphate* – crystalline, C.P.
3. *Sodium sulphate-mercuric oxide powder*: Mix 454 gm sodium sulphate and 24.4 gm mercuric oxide.
4. *Granulated zinc*
5. *Sodium hydroxide-sodium sulphide solution*: First, prepare approximately 16 N sodium hydroxide by dissolving 640 gm NaOH in water and diluting to 1 liter.
   Prepare sodium sulphide solution by dissolving 34 gm Na_2S in 85 ml of distilled water. Dilute to 1,000 ml with 16 N NaOH.
6. *Nessler reagent*: As in ammonia analysis.

Procedure

The method described here is essentially the Kjeldahl method for nitrogen analysis especially adapted for pond water analysis.

1. Shake water sample to bring all particulate matter into uniform suspension. Measure carefully 100 ml of water sample into a clean 500 ml Kjeldahl flask. Add 16 ml of concentrated sulphuric-salicylic acid solution and shake well. Allow to stands for at least 30 minutes.
2. Add 5 gm (1 teaspoon) of sodium thiosulphate crystals and shake well. Heat flask under hood until white fumes begin to evolve profusely. This takes about 45 minutes to one hour. Use asbestos shield under flask.
3. Add 5 gm (1 teaspoon) sodium sulphate-mercuric oxide powder. Continue heating sample until the liquid is clear. Continue digestion for another 30 minutes and cool.
4. Add 250 ml ammonia-free distilled water (page 286). Add ½ teaspoon granulated zinc. Alkalize with 60 ml NaOH–Na_2S solution. Do this by adding the alkali carefully along the side of flask to avoid violent reaction.
5. Set up Kjeldahl distillation apparatus, consisting of condenser, condenser bulb, receiving flask, and the Kjeldahl flask with the sample (page 289). Check that the fitting are airtight. Shake the sample to mix thoroughly when fitting is made.

6. Collect 255 ml of distillate in 20 ml 0.1 N HCl. This takes about 1 to 2 hours. Dilute to 300 ml.
7. Simultaneously with the sample analysis, subject 100 ml of nitrogen-free distilled water to the same treatment procedures as above to be used as reagent blank.
8. The total nitrogen in the original 100 ml sample is now all converted into ammonia in the 300 ml distillate. Nesslerize a 50 ml portion and evaluate concentration by using the spectrophotometer (ammonia analysis).
9. Multiply ppm N measured by nesserization by 3 to obtain ppm in sample.

Phosphates

This method is for the determination of phosphates in solution. It cannot be used in waters containing more than 0.2 ppm arsenates or arsenites as these ions give a blue colour similar to the phosphate ion.

Reagents

1. *Ammonium Molybdate-sulphuric acid solution*: Dissolve 25 gm ammonium molybdate in 200 ml of water heated to 60° C and filter. Pour 280 ml arsenic- and phosphorus-free concentrated sulphuric acid slowly and carefully into 420 ml distilled water. After both solutions have cooled, add the ammonium molybdate solution, slowly with shaking, to the sulphuric acid solution. Cool, dilute to 1 liter with distilled water. This is approximately a 10 N sulphuric acid solution containing approximately 2.5 gm of ammonium molybdate per 100 ml.
2. *Stannous chloride solution*: Dissolve 1 gm $SnCl_2 . 2H_2O$ in 5 ml of concentrated HCl. Dilute to 50 ml with distilled water. Filter into a 125 ml aspirator bottle and cover the liquid surface with white mineral oil. The aspirator bottle tubulature may be fitted with a rubber tube, drawn out glass tubing and pinchcock for drop-by-drop delivery of the solution. This solution deteriorates. When a blue colour fails to develop after its addition to the molybdate-treated phosphate solution, discard and make up fresh.

3. *Standard phosphate solution*: Dissolve 0.2195 gm of pure potassium-dihydrogen-phosphate (KH_2PO_4) and dilute to 1,000 ml with distilled water. This solution contains 50 ppm phosphorus and is too concentrated to use directly. A second stock solution is made by taking 50 ml of the first stock solution and diluting to 500 ml. This second stock solution contains 5 ppm P and is used for making the standard solutions for comparisons.

Procedure

1. Using standard phosphate (containing 5 ppm P), prepare standards containing 0.025, 0.05, 0.10, 0.25, 0.5, 0.75 and 1 ppm P as follows :

For Standards with Following ppm N	*Dilute Following Number of ml of the Standard Phosphate (Containing 5 ppm P) to 50 ml*
0.025	0.25
0.05	0.5
0.10	1.0
0.25	2.5
0.50	5.0
0.75	7.5
1.00	10.0

2. Add each standard 2 ml of the ammonium molybdate-sulphuric acid solution and mix by shaking.
3. Add 5 drops $SnCl_2$ solution and shake. A blue colour should develop promptly with the intensity of colour increasing with increase in phosphorus. If a blue colour does not develop, the stannous chloride solution has deteriorated and fresh one must be prepared.

   The blue colour begins to fade in 10 to 12 minutes and an additional drop of stannous chloride should be added to bring the full colour back.
4. Measure 50 ml of the water to be tested and place in a 100 ml Erlenmeyer flask.
5. Add 2 ml of the ammonium molybdate-sulphuric acid and mix by shaking.

6. Add 5 drops of $SnCl_2$ solution and shake.
7. Compare the colour produced by the phosphorus in the sample with that in the standards and estimate the ppm P.

The comparison can be made in 50 ml Nessler tubes or directly in the Erlenmeyer flask, or by use of the spectrophotometer as described on the following pages.

Procedure for making the Calibration Curve in the Spectrophotometric Method of Phosphorus Analysis

1. Switch on the "Spectronic 20 Colourimeter" for warm-up period of about 15 minutes before using. Be sure to use the 1P4O photocell and a red filter. Set the wavelength at 690 millimicrons. (Read the Reference Manual carefully before using the instrument).
2. Using the standard phosphate, prepare standards containing 0.025, 0.05, 0.10, 0.25, 0.50, 0.75 and 1.0 ppm P as described in Section 1 on page 298.
3. Add 2 ml ammonium molybdate-sulphuric acid to each standard.
4. Have 8 clean, dry cuvettes ready.
5. Place approximately 40 ml of the standard blank into one of the cuvettes (The blank here is 50 ml of distilled water as was used in preparing the standards, plus 2 ml ammonium molybdate plus 5 drops stannous chloride.)
6. Insert the cuvette into the sample holder of the spectrophotometer, being careful to set the index line on the cuvette opposite the index line on the sample holder.
7. Adjust the instrument to 100 per cent transmittance (or 0 optical density) with the blank.
8. Add 5 drops stannous chloride to each of the standards treated with molybdate in Section 3.
9. A blue colour should develop promptly which intensifies for a period and then slowly becomes less intense. Set up so that reading on the spectrophotometer are made just 10 minutes a after the stannous chloride is added. (If no colour develops, make up fresh stannous chloride).

10. Place approximately 40 ml of each standard in cuvettes and read per cent transmittance and optical density. Measure per cent transmittance and optical density of each standard successively and as quickly as possible.
11. Plot the optical (ordinate) against the corresponding concentration (abscissa) in standard graph paper. The curve a straight line passing through the origin, is the calibration curve.
12. To 50 ml of the test water, add 2 ml ammonium molybdate-sulphuric acid. Place 40 ml in a cuvette and reset the spectrophotometer at 100 per cent transmittance as in Section 7 above. This is the blank for the test water.
13. The each 50 ml sample, add 2 ml ammonium molybdate and 5 drops stannous chloride.
14. Place in a cuvette, read 10 minutes after the stannous chloride was added and compare to the curve for the standards (Section 11).

Total Hardness

Hardness of water is due principally to soluble salts of divalent cations, principally calcium and magnesium. Temporary hardness is due to calcium and magnesium bicarbonates. This is called 'temporary' because upon boiling, the bicarbonates change to carbonates and part of the calcium and magnesium are precipitated*.

$$Ca(HCO_3)_2 \rightarrow CaCO_3 + H_2O + CO_2$$

Permanent hardness is due principally to soluble Ca and Mg carbonates and salts of inorganic acids (*e.g.* $CaSO_4$). Permanent plus temporary hardness = total hardness.

(* Solubility at 25° C of $CaCO_3$ is 14 ppm; of $MgCO_3$ 106 ppm; of $CaSO_4$ 2090 ppm.)

Principle

The indicator forms a wine-red colour with Mg ions. A standard solution of the sodium salt of ethylendiamine tetraacetic acid (sometimes called versene) is used for titration. It combines first with free Ca ions, next with free Mg ions, and, finally at the end point extracts Mg from the indicator thus changing the colour from wine-red to blue.

The blue end-point occurs at pH 7.0-10 and a buffer solution must be used which keeps the pH at approximately 9-10. This is necessary because the indicator has two colour changes:

Wine-red $\longleftrightarrow$ blue $\longleftrightarrow$ orange

pH 6.3 pH 11.5

Reagents

These reagents can be purchased ready to use or made as follows:

1. *Buffer*: Mix 67.5 gm NH_4Cl with 570 ml concentrated NH_4OH. Dilute to 1 liter.
2. *Indicator*: Dissolve 4.5 gm hydroxylamine HCl and 0.5 gm of either Calmagite or Eriochrome black T (also known as Chrome black T or F241) in 70 per cent alcohol. Calmagite is more stable and is said to give a sharper end point. Eriochrome black T indicator should be replaced every 2 to 4 weeks.
3. *Standard $CaCl_2$*: Dissolve 1.00 gm pure $CaCO_3$ in about 10 ml diluted HCl (1·10) and diluted with distilled water to exactly 1 liter. Store in a glass-stoppered bottle. 1 ml = 1.0 mg $CaCO_3$.
4. *Standard sodium ethylendiamine tetraacetate (EDTA)*: Dissolve 4.0 gm disodium dihydrogen ethylendiamine tetraacetate (also known as disodium dihydrogen versenate) and 0.1 gm $MgCl_2 . 6H_2O$ in 750 ml distilled water and dilute 1 liter in a pyrex volumetric flask. Standardize by titrating against 25 ml of standard $CaCl_2$ solution plus 1 ml of buffer plus 4 drops indicator. Dilute so that 1 ml of standard = 1.0 mg $CaCO_3$.

Procedure

1. Pipette 50 ml water sample into a 250 ml Erlenmeyer flask.
2. Add 1 ml buffer and mix.
3. And 4 ml drops indicator.
4. Titrate with above standard solution. At the end point, the solution should clear and should change from wine-red to pure blue.

5. Multiply ml standard used x 20 = total hardness as ppm $CaCO_3$.
6. For very soft waters, 200 ml sample with 4 ml buffer may be titrated and the ml of standard required multiplied by 5 to obtain ppm $CaCO_3$.

Calcium in Fish Pond Waters

Reagents

1. *NaOH Solution*: Weigh approximately 40 gm pure NaOH and dilute to 1 liter with freshly boiled and cooled distilled water (approximately 1 N).
2. *Calcium indicator*: Mix 0.2 gm Murexide (also called NH_4 purpurate) with 100 gm NaCl. Grind to a uniform mix with a mortar and pestle. (Eriochrome blue black R may be substituted for Murexide; this gives a colour change from red to blue whereas, the Murexide end-point is orchid-purple and more difficult to detect.)
3. *Standard sodium ethylendiamine tetraacetate*: (See Total Hardness).

Procedure

1. Measure 50 ml water into a 250 ml Erlenmeyer flask.
2. Add 2 ml of NaOH solution and stir.
3. Add 0.2 calcium indicator and stir. (If Ca is present, colour is salmon pink with Murexide).
4. Titrate with standard sodium ethylendiamine tetraacetate, stirring constantly. Towards the end point, the colour has a purple tinge. At the end point, the colour is orchid-purple.
5. Multiply ml standard required x 20 Ca (as ppm $CaCO_3$).

Strontium and barium are included as calcium by this method, but are usually present in very small amounts.

Magnesium in Fish Pond Waters

Total hardness (as ppm $CaCO_3$) – Ca (as ppm $CaCO_3$) = Mg (as ppm $CaCO_3$).

Mg (as ppm $CaCO_3$) × 0.84253 = Mg (as ppm $MgCO_3$).

This is an estimate only as "hardness" may include other salts in addition to calcium and magnesium. American Public Health Association (APHA) lists cadmium, lead and zinc as included in hardness by EDTA titration.

Salinity

The average percentage compositions of dissolved solids in river waters and in sea waters are as follows:

Ion	*Average River Water*	*Sea Water*
CO_3^{--}	35.15 (HCO_3^-)	0.41
SO_4^{--}	12.14	7.68
Cl^-	5.68	55.04
NO_3^-	0.90	–
Ca	20.39	1.15
Mg	3.41	3.69
Na	5.79	30.62
K	2.12	1.10
$(Fe, Al)_2O_3$	2.75	–
SiO_2	11.65	–
Sr, H_3BO_3, Br^-	–	0.31
Total	100.00	100.00

The percentage composition of river waters varies widely, depending upon the soils, the kinds and amounts of pollution and the amount of rainfall on the watershed. Consequently, the total dissolved salt content of fresh waters is best determined by evaporation and incineration of a sample followed by weighing the incinerated salts. Incineration is necessary for the destruction of organic matter. It also converts carbonates and bicarbonates to oxides.

Salinity of sea water is defined as the total amount of salts in solution expressed as gm per kg of sea water, with all carbonates converted to oxides, all bromides and iodides expressed as chlorides and all organic mater completely oxidized.

In the oceans the percentage composition of dissolved solids is relatively constant. Consequently, it was found possible to relate the chloride content to the total dissolved salt content, or salinity, of sea water.

The relationship is : Salinity = 0.03 + 1.805 × chlorinity.

Chlorinity is the total amount of chloride present in 1 kg of sea water, plus all bromides and iodides expressed as chlorides. Chlorinity is thus determined by titration of sea water with standard $AgNO_3$, using a few drops of K_2CrO_4 as indicator.

The salinity of brackishwaters, where the sea water and river waters are mixing, is also commonly determined by the silver nitrate titration. This is not exactly accurate because in the zone of mixing, the percentage composition of salts in solution is not constant and often differ materially from that is sea water, particularly in areas of low salinity. However, for practical purpose, it appears sufficiently accurate to validate its use biologically.

The reaction is : $NaCl + AgNO_3 \rightarrow AgCl + NaHO_3$

As long as chlorides remain in solution the silver is used to form insoluble AgCl. As soon as the last soluble chloride is precipitated, the following reaction occurs with the potassium chromate.

$$K_2CrO_4 + 2Ag\,NO_3 \rightarrow Ag_2CrO_4 + 2KNO_3$$

The Ag_2CrO_4 precipitate has a red colour.

Reagents

1. *0.1595 N silver nitrate*: Weigh 27.096 gm C.P. $AgNO_3$ crystals and dilute with distilled water to 1 liter in a volumetric flask. Keep in a dark place.
2. *5 per cent potassium chromate indicator**: Dissolve 5 gm of K_2CrO_4 in 100 ml distilled water.

   (* Sodium fluorescinate (also called uranin) is reported to give a more accurate end-point than potassium chromate. The precipitate changes from greenish yellow to pink at the end point.)

Procedure

1. Measure 10 ml of the water to be analyzed into a 125 ml Erlenmeyer flask.
2. Add 5 drops potassium chromate indicator. Use a magnetic stirrer.

3. Titrate with standard 0.1595 N silver nitrate to a permanent pink.
4. Express salinity as parts per thousand ($^0/_{00}$). Each 1 m of 0.1595 N silver nitrate = 1.0 $^0/_{00}$ or 1 ppm salinity.

For more accurate expression of salinity, the above figure for salinity should be corrected by the following amounts:

Salinity Found	*Correction to be Applied*
8	+0.15
10	+0.16
12	+0.19
14	+0.20
16	+0.23
20	+0.23
24	+0.20

Chlorides

The determination of salinity is essentially the determination of chlorides plus bromides and iodides by precipitation with silver nitrate. Since bromides and iodides are present only as traces in most fresh waters, the silver titration method may be used to determine the chlorides*.

(* In polluted waters containing soluble proteins, this method will give inaccurate results because silver will cambine with proteins.)

Procedure

1. Measure 500 ml of the water to be analyzed into a 1000 ml Erlenmeyer flask.
2. Add 5 drops potassium chromate indicator. Use a magnetic stirrer.
3. Titrate with standard 0.1595 N silver nitrate to a permanent pink.
4. Each 1 ml of 0.1595 N silver nitrate = 5.65 mg chloride ion (Cl^-), or

   = 9.3 mg chlorides as NaCl.

5. To express as Chloride (Cl^-) in sample:

   ml 0.1595 N silver nitrate titre × 5.65 × 2 = ppm Cl^-

   As NaCl equivalent in sample:

   ml 0.1595 N silver nitrate titre × 9.3 × 2 = ppm as NaCl.

Chlorine

O-Tolidine Method for Free and Combined Available Chlorine.

City water suppliers are usually chlorinated for the purpose of disinfection and contain from traces to 2 ppm residual chlorine. Ponds are occasionally chlorinated to control diseases and parasites. Part of the chlorine combines with organic matter and part slowly reacts with water, especially when exposed to sunlight.

$$2\,Cl + H_2O \rightarrow HOCl + HCl$$

$$4\,HOCl \rightarrow 4\,HCl + 4\,O$$

or

$$4\,HOCl + 4\,HCl \rightarrow 4\,H_2O + 8\,Cl$$

$$8\,Cl + 4\,H_2O \rightarrow 8\,HCl + 4\,O$$

Free available chlorine is in the form Cl, OCl^- or HOCl; combined available chorine is present in such form as the chloramines, which are often used for city water disinfection by reaction of NH_4OH and HOCl, or they may be formed in natural waters:

$$NH_4OH + HOCl \rightarrow NH_2Cl + 2H_2O$$ (Only monochloramine formed above pH = 8.5)

$$NH_4OH + 2\,HOCl \rightarrow NHCl_2 + 3H_2O$$ (Only dichloramine found when pH = 4.4 – 5.0)

At pH 5.0 to 8.0, mono- and dichloramine occur together.

Residual chlorine is that in free available forms and in combined available forms. The orthotolidine method measures total available residual chlorine. This reagent reacts to give a yellow colour, with the intensity dependent upon the amount of chlorine. The pH during the reaction should be 1.3 or less.

Waters containing more than 0.3 ppm, Fe, 0.01 ppm, Mn (ic) or over 0.1 ppm N (as NO_2^-) give interfering colour. Many pond waters low in oxygen contain sufficient iron to give a yellow colour with O-tolidine in the absence of available chlorine.

Reagents

1. *Orthotolidine reagent*: Dissolve 1.4 gm orthotolidine dihydrochloride in 500 ml distilled water. Add with stirring to 500 ml of dilute HCl (1-3). Store in glass-stoppered amber bottles in the dark. This solution should be prepared fresh every 5 to 6 months.
2. *Permanent colour standards for chlorine estimation*:
   a. *Stock phosphate buffer*: Dissolve 11.43 gm anhydrous disodium phosphate (Na_2HPO_4) plus 23.08 gm anhydrous potassium dihydrogen orthophosphate (KH_2PO_4) in 1 liter water. Allow to stand 48 hours before using.
   b. *Phosphate buffer 0.1 M*: Filter the above stock phosphate buffer and dilute 200 ml to 1 liter with distilled water. This gives a pH of 6.4.
   c. K chromate – dichromate standard : Dissolve 0.155 gm $K_2Cr_2O_7$ and 0.465 gm K_2CrO_4 with 0.1 m phosphate buffer and dilute to 1 liter with this same buffer. The colour is equiavalent to 1 ppm available chlorine by the O-tolidine method. Lower standards can be obtained by dilution as follows:

Equivalent to ppm Cl	*ml standard $Cr_2O_7 - CrO_4$*	*Diluted with 0.1 m Buffer ml*
1.0	10	0
0.5	5	5
0.3	3	7
0.2	2	8
0.1	1	9

   d. *Arsenite reagent*: Dissolve 5 gm sodium meta arsenite ($NaAsO_2$) in 1 liter distilled water.

Procedure

1. Collect in a glass vial 10 ml sample of water to be tested.

2. Add 0.5 ml O-tolidine reagent. Compare the colour with permanent standards.

For field work, or a quick check, the development of even a light yellow colour is indicative of sufficient chlorine to be injurious or toxic to fishes.

Alternate Method where Fe, Mn, or NO_2 Interfere

1. Add 10 ml of the water sample to 0.5 ml O-tolidine reagent. Mix quickly and in exactly 5 minutes compare with colour standards. This represents total residual chlorine plus interfering colours (T).
2. To 0.5 ml arsenite reagent add 10 ml of the water sample. Mix quickly, and rapidly add 0.5 O-tolidine. Mix and compare with colour standards in exactly 5 minutes. This represents the interfering colours (I).
3. The difference, T – I = total residual chlorine.

Titration Method for Free Chlorine

This method is based on the reaction:

$$Cl_2 + 2\,Na_2S_2O_3 \rightarrow Na_2S_4O_6 + 2NaCl$$

The indicator used is O-tolidine.

Reagents

1. O-tolidine reagent as prepared on page 307.
2. 0.025 N sodium thiosulphate as prepared for oxygen analysis on page 263.

Procedure

1. To 1,000 ml water sample, add 2 ml O-tolidine reagent to develop a yellow colour.
2. Titrate with 0.025 N sodium thiosulphate to disappearance of the yellow colour.
3. Multiply ml of titre by 0.88 to obtain ppm available chlorine.

Calculation

1 ml 0.025 N sodium thiosulphate = 1 ml 0.025 N Cl

1N Cl = 35.46 gm/liter

0.025 N Cl = 35.46 × 0.025 = 0.8865 gm/liter

1 ml 0.025N Cl = 0.8865 mg Cl

Arsenic

Arsenicals are used in ponds for weed control at rates ranging from 2 to 8 ppm As_2O_3. The chemical most commonly used is sodium arsenite, approximately $NaAsO_2$, in a aqueous solution containing either 1.8 or 3.6 kg (1 lb = 0.4536kg) As_2O_3 per 3.785 litres.

After application to the water, the arsenic is rapidly absorbed by the aquatic plants, including plankton, and by the bottom soils. Applications of As_2O_3 in excess of 8 ppm reduced fish production in ponds.

The method here employed is the Gutzeit test because it is specific for arsenic. The colourimetric method give for phosphorous (molybdate-stannous chloride) will also yield a blue colour with arsenic and it is occasionally necessary to determine arsenic in order to be sure that it is not being falsely reported as phosphorus.

Arsenic present in aqueous solution can be determined directly. However, that in organic matter must first have the organic matter destroyed. This is done by placing the weighted sample in a Kjeldahl flask, adding 2 to 8 ml concentrated H_2SO_4 and 20 ml concentrated nitric acid. This is boiled to about 15 ml, add perchloric acid and nitric acid drop by drop until the solution clears. The digested sample is washed into a volumetric flask, diluted to a known volume and aliquots taken for arsenic analysis.

Gutzeit Analysis for Arsenic

Reagents

1. *Dilute HCl*: Dilute 1 part special arsenic-free HCl with 6 parts distilled water.
2. *Zinc*: Use 20 mesh metallic zinc, arsenic free.
3. *Lead acetate solution*: Dissolve 10 gm $Pb(CH_3COO)_2.3H_2O$ in 100 ml distilled water.
4. *Lead acetate cotton*: Soak 2-inch length of either absorbent cotton or dentist's roll cotton in the lead acetate solution. Wring out excess moisture and store moist cotton in a wide-mouth bottle.

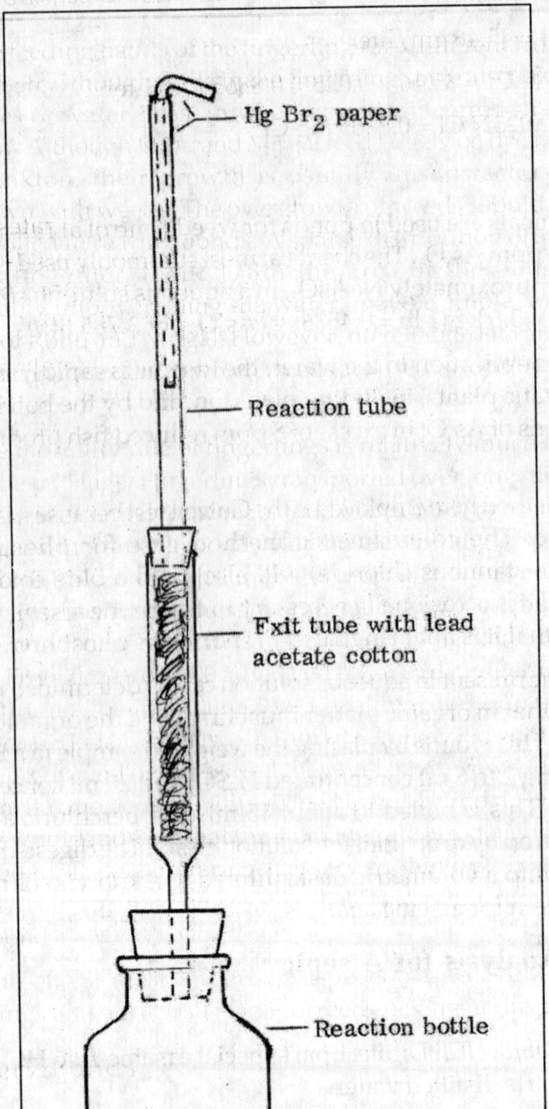

Figure 10.6: Gutzeit Arsenic Bottle

5. *Mercuric bromide solution*: Dissolve 5 gm $HgBr_2$ in 100 ml 95 per cent alcohol.
6. *Mercuric bromide paper*: Purchase commercially prepared Gutzeit arsenic papers (or cut strips to fit the Gutzeit reaction tube from filter paper). Soak in the mercuric bromide solution and dry by draining off excess fluid and waving papers in the air. Store strips after drying in a desiccator placed in the dark.
7. *As_2O_3 standard*: Dissolve 5.00 gm arsenious oxide (As_2O_3) in 25 ml 10 per cent HCl and dilute to 1 liter. 1 ml = 5 mg As_2O_3.

Procedure

1. Place lead acetate cotton in the large exit tube as illustrated.
2. Insert a strip of mercuric bromide paper in the reaction tube as illustrated.
3. Place 3 gm metallic zinc in the reaction bottle.
4. Add 15 ml dilute HCl (1-6) to the reaction bottle.
5. Add the aliquot sample and insert the exit tube assembly as illustrated. Be sure to position the reaction tube in a vertical position so that the bromide paper will be coloured evenly on both sides.
6. Allow the reaction, which releases arsine, AsH_3, along with the nascent hydrogen, to proceed at room temperature. The arsine reacts with mercuric bromide paper to give an orange to yellow stain on the paper.
7. After 1 to 2 hours, when effervescence has ceased, remove the $HgBr_2$ strips. Measure the stain to the nearest mm on both sides of the paper and compare the average with a graph prepared from known solutions.
8. The standard graph is prepared from the standard solution at 5 mg intervals from zero to 30 mg As_2O_3.

Sodium, Potassium and Calcium by Flame Photometer

Sodium, potassium and calcium each produce a colour (yellow, red-violet and yellow-red, respectively) when burned in a flame. In the flame photometer, the solution containing the sample is atomized into a oxygen-gas flame. A filter for the particular element is inserted to allow only passage of the desired wavelength, if light and the intensity of colour measured by a photometer. The intensity is related to the concentration of the element present in the solution.

Since flame photometers vary greatly in exact method of operation, be sure to read the operating directions for the particular model you are to use.

Reagents and Standards

1. *1 per cent Sterox stock*: Weigh 5 gm Sterox into a 200 ml beaker, add 200 ml distilled water and stir until dissolve. Transfer to 500 ml pyrex volumetric flask and dilute to the mark with distilled water.
2. *0.02 per cent Sterox atomizing solution*: Dilute 20 ml of 1 per cent Sterox stock in a 1,000 ml pyrex volumetric flask to 1,000 ml with distilled water.
3. *500 ppm Na Standard*: Dry reagent grade NaCl by heating for 30 minutes at 125° C. Cool in a desiccator. Place 1.271 gm in a 1,000 ml volumetric pyrex flask, add 900 ml distilled water, shake until dissolve and fill to mark.

   Prepare standards as follows :

To Obtain ppm Na	*Add 1 ml of 1 per cent Sterox Plus Following ml of 500 ppm Na and Dilute to 50 ml*
1	0.1
5	0.5
10	1.0
25	2.5
50	5.0
100	10.0

4. *500 ppm K Standard*: Dry reagent grade KCl for 30 minutes at 125°C and cool in a desiccator. Weigh 0.9535 gm and place in a 1,000 ml pyrex volumetric flask. Dissolve and fill to the marks with distilled water.

   Make standards by dilution with distilled water as follows:

To Obtain ppm K	*Add 1 ml of 1 per cent Sterox Plus Following ml of 500 ppm K and Dilute to 50 ml*
1	0.1
5	0.5
10	1.0
25	2.5
50	5.0
100	10.0

5. *500 ppm $CaCO_3$ Standard** : Dry reagent grade $CaCO_3$ at 150° C for 30 minutes and cool in a desiccator over soda-lime. **Weigh 0.500 gm and place in a 1,000 ml pyrex volumetric flask. Add 25 ml dilute HCl (1:1) and swirl until calcium is dissolved. Dilute to 1,000 ml with distilled water.

   Make standards as follows:

To Obtain ppm $CaCO_3$	*Equivalent to ppm Ca*	*Add 1 ml of 1 per cent Sterox Plus Following ml 500 $CaCO_3$* and Dilute to 50 ml*
1	0.4	0.1
5	2.0	0.5
10	4.0	1.0
25	10.0	2.5
50	20.0	5.0
100	40.0	10.0

(* The 1,000 ppm $CaCO_3$ standard prepared for the determination of hardness may be used here and would require the dilution of the ml given in the right column of the above table to 100 ml instead of 50 ml to obtain the ppm given in the left column.)

(# A mixture of $NaOH^-$ CaO with 18 per cent water. Commercially available as white granules that absorb CO_2)

6. *Standard Na-K-Ca check solution*: After the calibration curves are obtained, a 50 ppm and a 100 ppm Na-K-$CaCO_3$ check solution can be used to set the galvanometer scale. These are prepared as follows:

a. *25 ppm Na-K-CaCO$_3$*: Pipette accurately into a 250 ml pyrex volumetric flask,
12.5 ml standard 500 ppm Na,
12.5 ml standard 500 ppm K,
12.5 ml standard 500 ppm CaCO$_3$, and
50 ml 1 per cent Sterox.
Dilute to 250 ml with distilled water.

b. *100 ppm Na-K-CaCO$_3$*: Pipette accurately into a 250 ml pyrex volumetric flask,
50 ml 1 per cent Sterox,
50 ml standard 500 ppm Na,
50 ml standard 500 ppm K and
50 ml standard 500 ppm CaCO$_3$.
Dilute to 250 ml with distilled water.

Preliminary Steps (Coleman Flame Photometer)

1. Plug in line cord, Allow instrument to warm up at least 15 minutes before use. Where many tests are to be run, allow to run continually, turning off only the burner between tests.
2. To light flame (for natural gas) :
   a. Light match.
   b. Turn of gas full, hold match inside top edge of cylindrical screen by reaching through the door opening.
   c. When yellow flame appears at top of chimney, bring oxygen pressure to about 2.3 kg per square inch.
   d. Check flame, if concentric, increase oxygen pressure to 5.9 kg per square inch. (If flame is not concentric consult book of directions of cleaning the clogged capillary).
   e. To shut off flame, first turn of oxygen and immediately turn off gas.
3. Place filter for Na, K, or Ca in slot, name side to front.

Preparing Calibration Curves

1. Set galvanometer coarse control almost to full counter-clockwise position and galvanometer fine control to center, control knob in form of filter to 10.
2. Place beaker of blank solution (0.02 per cent Sterox in distilled water) in lifter, close door and atomize. Set galvanometer reading to zero with blank control.
3. Pour standards 1,5, 10, 25, 50 and 100 ppm into 10 ml pyrex beakers. Prepare separately the above for Na, K and $CaCO_3$.
4. Open door, replace beaker with 25 ppm Na, set galvanometer index to 65.
5. Flush atomizer with atomizing solution (0.25 per cent Sterox) and repeat step 2.
6. Repeat step 4 and set the galvanometer to exactly 65.
7. Insert in turn the standards prepared in step 3, reading and recording each galvanometer reading for each standard from 1 to 50 ppm.
8. Clean atomizer with atomizing solution.
9. Repeat steps 4 through 9 for Na, K and $CaCO_3$ standards.
10. Prepare a calibration curve for galvanometer reading and ppm from 1 to 50 for Na, K and $CaCO_3$.
11. For higher concentration, repeat steps 4 through 6 with 100 ppm set a 65 on the galvanometer and through steps 7 and 8 with 50 ppm. Prepare a calibration curve for 50-100 ppm.

Procedure

1. Plug in instrument and light flame as described above.
2. Filter water samples through Whatman No. 42 filter paper.
3. Insert Na filter.
4. Clean atomizer with atomizing solution.
5. Set galvanometer at 65 with 25 ppm Na-K-$CaCO_3$ check solution for low range; or, set at 65 with 100 ppm check solution for high range.

6. Place beaker containing sample in lifter and atomize. Read galvanometer and refer to calibration curve for ppm.
7. Clean atomizer with atomizing solution.
8. Repeat steps 6 and 7 for each sample.
9. Remove Na filter and insert K filter and repeat steps 5 through 7.
10. Remove K filter and insert Ca filter and repeat steps 5 through 7.

Dichromate Oxidation Analysis for Organic Matter

Chemical Oxygen Demand (COD)

Total organic matter in water may be determined by evaporation to dryness, followed by determination of the loss of weight upon incineration. It may also be determined indirectly by chemical oxidation and the relationship between the average amount of oxygen required for oxidation of organic materials. One method currently in use is that for the determination of COD by oxidation with potassium permanganate. Dichromate oxidation appears superior because the reagent is not subject to auto-decomposition as is true of the permanganate, and the oxidation is more complete. Permanganate oxidizes only 10 to 54 per cent of organic compounds as compared to practically 100 per cent by the dichromate method.

Boiling $K_2Cr_2O_7$ and sulphuric acid oxidizes most organic compounds with the quantitative reduction of Cr_2O_7 = to Cr^{+++}. The excess $K_2Cr_2O_7$ can be titrated with ferrous ammonium sulphate. The reaction is:

$$6\,Fe(NH_4)_2(SO_4)_2 + K_2Cr_2O_7 + 7H_2SO_4 \rightarrow K_2SO_4 + 3Fe_2(SO_4)_3 + Cr_2(SO_4)_3 + 6(NH_4)_2SO_4 + 7H_2O$$

Interfering Substances

Aromatic hydrocarbons and pyridine are not oxidized appreciably. Volatile compounds may be driven off the boiling before being oxidized. Chloride interferes by combining with the silver catalyst. Nitrites and Fe^{++} give high results by reducing $K_2Cr_2O_7$.

The following method is adapted from APHA Standard methods.

Reagents

1. *Standard 1.000 N $K_2Cr_2O_7$ stock solution*: Dry reagent grade $K_2Cr_2O_7$ at 103° C for 2 hours. Cool in desiccators. Dissolve 49.037 gm in distilled H_2O in 1 liter volumetric flask. Dilute to 1 liter. This is stable for several years if kept tightly stoppered.
2. *Standard 0.025 N $K_2Cr_2O_7$*: Dilute exactly 25.0 ml of the above standard to 1 liter in volumetric flask, and 0.1 gm sulphuric acid to eliminate interference from nitrites.
3. *Sulphuric acid–silver sulphate reagent*: Dissolve 22 gm Ag_2SO_4 in a bottle of conc. H_2SO_4 (approximately 1 gm per 100 ml).
4. *Ferrous ammonium sulphate*: Dissolve 9.8 gm $Fe(NH_4)_2(SO_4)_2 \cdot 6H_2O$ in distilled water, add 20 ml conc. H_2SO_4, cool, and dilute to 1 liter with distilled water. Store in brown bottle. Standardize whenever used.
5. *Mercuric sulphate*: Reagent grade crystals (needed only if chlorides are 200 ppm or more in sample).
6. *Ferroin indicator*: This reagent may be purchased or prepared as follows : dissolve 1.5 gm 1, 10-phenanthroline monohydrate and 0.7 gm $FeSO_4.7H_2O$ in 100 ml distilled H_2O.

Special Apparatus

A Reflux apparatus using 125 or 250 ml flask with ground glass neck and Leibig or West condensers with ground glass joints must be used. Rubber or cork stoppers cannot be used since the rubber or cork will be oxidized by the dichromate. A hot plate should be used to heat the flasks.

Procedure

1. Add 20.0 ml of sample, 0.4 gm $HgSO_4$ (if needed), 10.0 ml 0.025 N $K_2Cr_2O_7$ and several glass boiling beads to reflux flask. Mix well.
2. Add carefully 30 ml sulphuric acid-silver sulphate reagent. Mix well. If the mixture is not thoroughly mixed at this point, the application of heat for refluxing may cause the mixture to be blown out the top of the condenser.
3. Reflux for 2 hours.

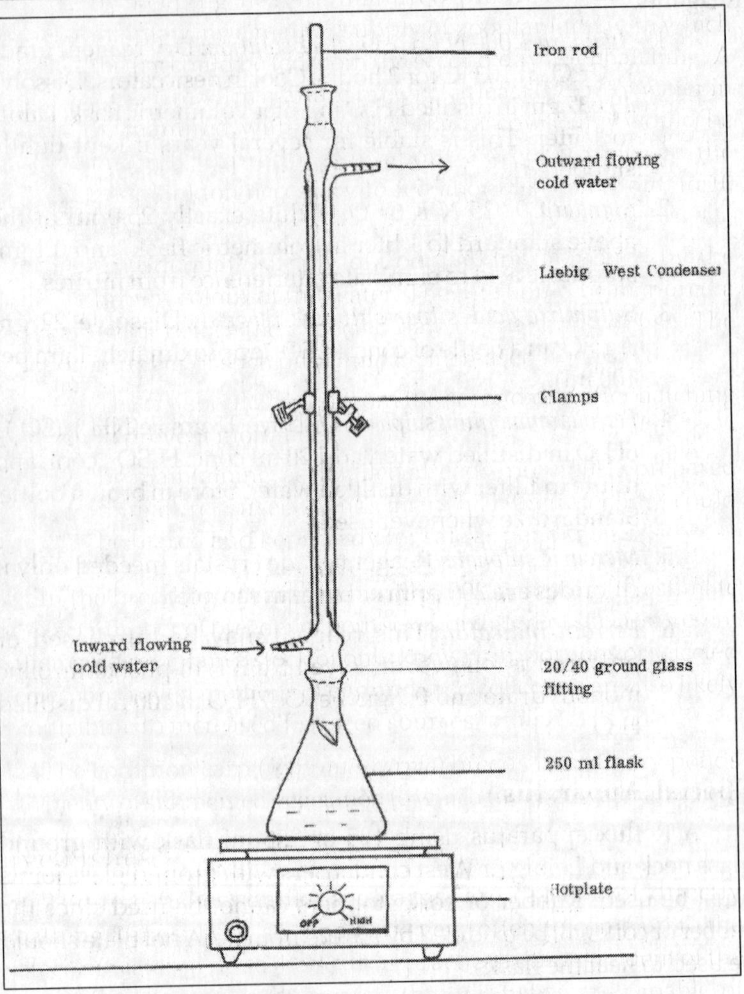

Figure 10.7: Reflux Apparatus for Chemical Oxygen Demand

4. Cool, wash down condenser with 20-30 ml distilled water. Dilute to 75 to 100 ml.
5. Titrate excess dichromate with ferrous ammonium sulphate using 2-5 drops ferroin indicator. Colour change is from blue-green to red-brown.
6. Carry 2 or more blanks (20 ml distilled water instead of sample) through steps 1 through 5 along with samples.

Standardization of Ferrous Ammonium Sulphate

1. Dilute 10.0 ml 0.025 N $K_2Cr_2O_7$ to about 100 ml and add 30 ml conc. H_2SO_4, allow to cool.
2. Titrate with ferrous ammonium sulphate, using 2 to 5 drops ferroin.

$$\text{Normality} = \frac{\text{ml dichromate} \times \text{normality of dichromate}}{\text{ml ferrous ammonium sulphate}}$$

Standardization should be performed in duplicate.

Calculation

$$\text{ppm COD} = \frac{(a-b)\, c \times 8000}{\text{sample vol. in ml}}$$

where,

a: ml $Fe(NH_4)_2(SO_4)_2$ for blank,

b: ml $Fe(NH_4)_2(SO_4)_2$ for sample, and

c: normality of $Fe(NH_4)_2(SO_4)_2$.

ppm organic matter = ppm COD × 0.7

The factor 0.7 is an empirically derived average value for organic matter.

Notes

1. The glassware used must be especially clean, even a tract of organic matter can cause errors.
2. For more concentrated samples, the sample may be diluted or the $K_2Cr_2O_7$ in the reflux flask may be increased in strength. The strength of the ferrous ammonium sulphate may be increased as well to keep the titration volume within a reasonable size.
3. The normality of the $K_2Cr_2O_7$ in the reflux flask does not need to be known since it does not appear in the calculations. However, the volume and normality of the dichromate must be the same in the sample and in the blanks.

Analysis of Fish Feeds

The elements of interest in fish feeds are protein, fat, digestible carbohydrate, indigestible fiber, and minerals. Since all determinations should be referred to the dry weight of the sample, the moisture content must also be determined. Protein, fat (ether extract), fiber and minerals (ash) are determined directly, the remainder (protein free extract) is considered to be a measure of digestible carbohydrate, and is usually determined by difference.

Sample Preparation

The moisture content may be determined on fresh plant materials (*e.g.*, leaves) with no preparation. Grind the dried materials from the moisture determination and use it for other determinations.

Quick-freeze fresh meat or fish samples. Within 24 hours, chop into small pieces while still frozen. Determine moisture content of frozen material. Grind the dried material and use it for other determinations.

Grind dry feeds (*e.g.*, grains, pelleted feeds) to pass through a 20 mesh per inch screen. This material may be used for all determinations or the dried material from the moisture determination may be used for other determinations.

Moisture Content

Procedure

1. Heat a clean porcelain crucible or dish at 100°C for 2 hours. Cool in desiccator and weigh accurately.
2. Weigh in this crucible or dish a sample of suitable size, 1 to 2 gm (or larger to provide dried material for other determinations, if desired).
3. Dry at 100° C for 10 hours. Cool in desiccator and weigh. Save for ash (or other) determination.

Calculation

$$\text{Per cent moisture} = \frac{100 \times \text{sample wt - dry wt}}{\text{sample wt.}}$$

If the dry residue is not used as the sample in another determination, the sample weight in that determination must be corrected for moisture. Multiply the sample weight by a factor (F) to get the dry weight of that sample.

$$F = \frac{100 - \%\ \text{moisture}}{100}$$

Total Ash Content

Procedure

1. Heat the weighted sample from step 3 of the moisture content procedure at 550°C for 4 hours.
2. Cool in desiccator and weigh.

Calculation

$$\text{Per cent ash} = \frac{\text{ash weight}}{\text{sample weight}} \times 100$$

Fat Determination (Ether Extract)

Reagents

1. *Anhydrous ether*: This should be stored in tightly closed container, preferable in a refrigerator. Caution – highly flammable.

Procedure

1. Carefully weigh 1 to 2 gm sample of material of known moisture content and place in alundum or fiber extraction thimble*. Insert thimble into glass tube.
2. Add 35 ml anhydrous ether to a weighed beaker and attach to Goldfisch extractor.
3. Place tube containing thimble into the Goldfisch extractor. Turn on cooling water. Raise the heating elements and heat at high temperature for 4 hours. Ether should boil, but not vigorously. Turn off heater and cool.
4. Remove thimble and replace with reclaiming tube. Turn on heater and distill over most of the ether leaving fat in the beaker. Remove reclaiming tube from extractor. Turn heaters to low and let rest of ether evaporate with tilted beaker-holders down on heating elements. Too high heat may char sample.

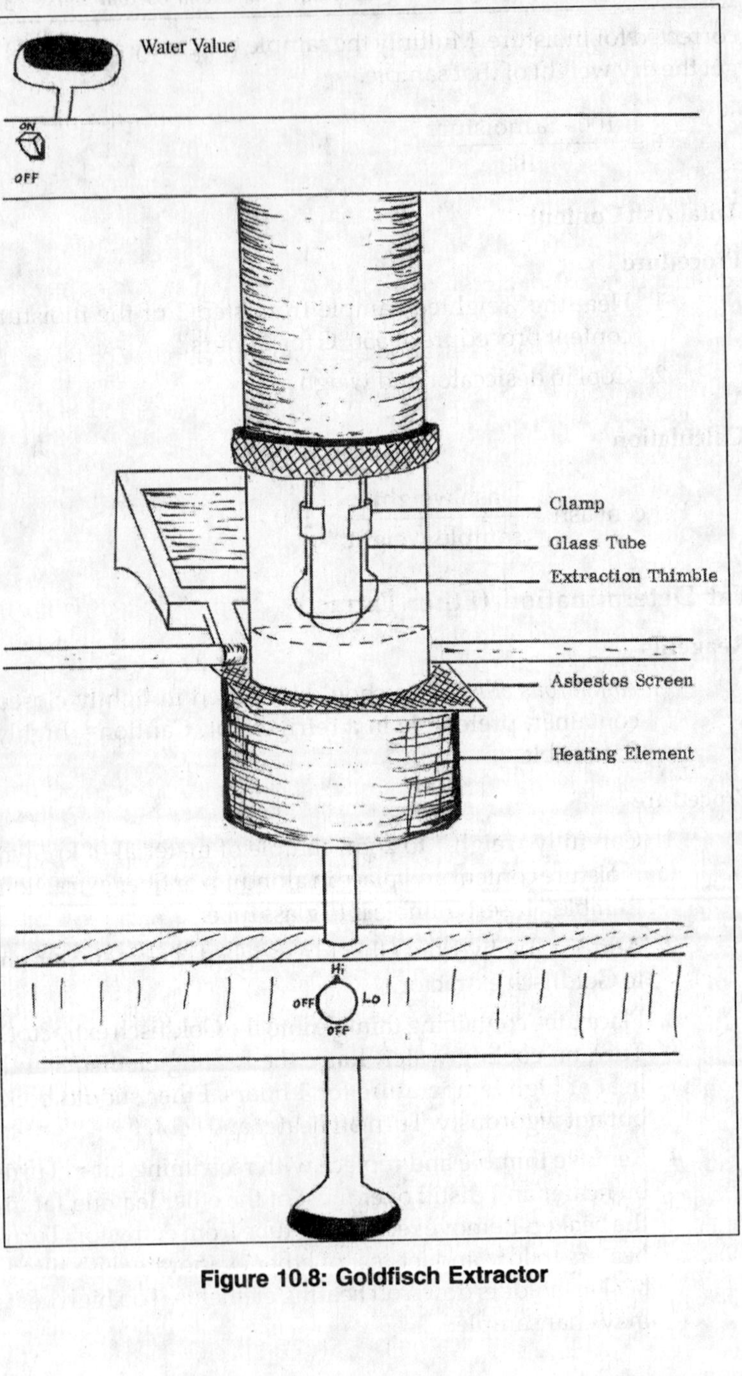

Figure 10.8: Goldfisch Extractor

5. Put beaker in oven at 60 – 80° C and dry for 1 hour. Cool in desiccator and weigh.
6. Save fat-free material in thimble for crude fiber or ash determination.

   (* May use fibre extraction thimble if sample is not to be ashed after ether extraction. Sample can be ashed in alundum thimble following ether extraction.)

Calculation

$$\text{Per cent Fat} = \frac{(\text{beaker} + \text{fat}) - \text{weighed beaker}}{\text{Dry weight of sample}} \times 100$$

Crude Fiber

The fiber measured by this procedure is an approximation of the indigestible matter in the feed. Accuracy and reproducibility depend on close adherence to the prescribed concentrations and timing.

Reagents

1. *Anhydrous ether*: As in fat determination.
2. *H_2SO_4, 1.25 per cent*: Add 7.0 ml conc. H_2SO_4 to 500 ml distilled water, mix and dilute to 1 liter.
3. *NaOH, 1.25 per cent*: Dissolve 12.5 gm NaOH in distilled water and dilute to 1 liter.
4. *Antifoam*: This may be purchased (Dow-Corning Antifoam A is recommended) or prepared as follows. Mix 1 part paraffin oil and 1 part octyl alcohol.
5. *Ethanol 95 per cent.*

Procedure

1. Extract with ether to remove fat or use sample remaining after fat extraction.
2. Warm sample gently in hood to dive off most of the remaining ether, then dry in oven at 100° C. If sample has previously been dried for moisture determination, dry for 1 hour to drive off remaining ether, if not, dry for 10 hours to drive of moisture as well. Cool in desiccator.

3. Weigh accurately an approximately 2 gm sample. Place in 600 ml beaker.
4. Add 200 ml boiling 1.25 per cent H_2SO_4 and 1 small drop antifoam (do not spray). Add boiling chips or beads.
5. Place beaker of hot solution on preadjusted digestion rack and boil for 30 minutes rotating beaker periodically to keep particles from adhering to sides.
6. Pour into 200 ml centrifuge tube and centrifuge for 5 minutes at 1800 rpm. Decant carefully. Add boiling water to tube until 2/3 full and stir or shake. Centrifuge and decant carefully. Repeat washing with water 3 more times.
7. Return residue to 600 ml beaker and add 200 ml boiling 1.25 per cent NaOH and boil for 30 minutes.
8. Transfer to centrifuge tube and centrifuge and wash with boiling water as in step 6.
9. Wash the residue from centrifuge tube into 25 ml alundum crucible with boiling (80°C) ethanol, 95 per cent. Filter by suction. Be sure all residue is in the crucible.
10. Allow the crucible to dry under a hood for a few minutes and then dry for 1 hour in a 100° C oven. Remove, cool in desiccator, weigh to get the weight of the fiber plus ash.
11. Replace crucible in muffle furnace for 3 to 5 hours at 550 to 600° C, remove, cool in desiccator, weigh to get weight of ash.

Calculation

$$\text{Per cent fiber} = \frac{b - c}{aF} \times 100$$

where,

a: Sample weight from step 3,

b: Weight or crucible plus sample from step 10,

c: Weight of crucible plus ash from step 11 and

F: Factor to correct for weight of extracted fat

$$F = \frac{100}{100\,\%\ \text{fat}}$$

Protein Nitrogen

The feed is digested in a Kjeldahl flask, transforming organic nitrogen to ammonium. The ammonia is distilled into a standard acid and determined by titration of the residual acid.

Reagents

1. *Potassium sulphate crystals.*
2. *Metallic mercury.*
3. *Concentrated H_2SO_4.*
4. *NaOH – Na thiosulphate reagent*: As in organic nitrogen analysis (page 292)
5. *Metallic zinc granular*
6. *4 per cent boric acid*: Dissolve 40 gm boric acid in 1 liter.
7. *Indicator solution*: Dissolve 0.5 gm bromcresol green and 0.1 gm methyl red in 100 ml 95 per cent alcohol.
8. *Standard 1/14 N HCl*: Prepared a stock constant boiling mixture as prepared for alkalinity test. Dilute to 1/14 N. The volume to be diluted to 1 liter is :

$$\text{ml stock solution} = \frac{1/14 \times 1000}{\text{normality of stock solution}}$$

Normality of HCl may be reduced to as low as 0.01 N for samples of low protein content.

Procedure

When feeds are to be analyzed, the following sample sizes should be taken:

1. If a protein content of over 20 per cent is suspected, weigh 0.7 gm samples.
2. If a protein content of under 20 per cent is suspected, weigh 1.4 gm samples.

A reagent blank containing no sample must be analyzed at the same time.

Procedure

1. Weigh dried sample and place in 800 ml Kjeldahl flask.

2. Add 10 gm K_2SO_4, 1 drop metallic mercury and 25 ml conc. H_2SO_4.
3. Digest under hood until contents of flask are clear. There will be a black residue formed on the sides of the flask. Swirl to wash residue down and allow it be digested.
4. Allow flask to cool. Add 230 ml of distilled water. Care should be taken to add the water slowly to keep the acid from "coughing" out of the flask. Swirl the flask to make K_2SO_4 crystals go back into solution. Allow to cool to room temperature.
5. Add 130 ml of NaOH-Na thiosulphate solution, several glass beads, and approximately 2 ml of granular zinc. Swirl to mix and distill.
6. Collect about 200 to 250 ml distillated in 25 ml of boric acid with 4 drops of indicator solution added. Condenser tip should be ¼ inch below surface of boric acid.
7. Titrate with 1/14 N HCl to end point (faint yellow colour).

Calculation

$$\text{Per cent N} = \frac{100\,(S - B) \times 14 \times \text{Normality of HCl}}{\text{sample wt. gm} \times 1000}$$

where,

S: ml HCl to titrate sample, and

B: ml HCl to titrate blank.

Per cent protein = per cent N × 6.25

Preparation of Soil for Analysis

Pond bottom soils to be analyzed for calcium, magnesium, inorganic phosphorus, pH, and organic matter are dried at a temperature of approximately 40° C. Drying may be done inside an oven or under infrared lamp. To facilitate grinding, soils should be broken up into small pieces as they dry up while it will be moist and soft. After drying, grind soils in a mortar and sieve through a 20-mesh screen. It takes 2 to 3 days to dry 1 pint of soil sample in a suitable-sized bowl.

Soil pH

Shake 20 gm of soil sample with 20 ml distilled water for 5 minutes. Let stand 30 minutes to 1 hour. Measure pH by means of a glass electrode pH meter. While the electrodes are immersed in the mixture, swirl the mixture by rotating the container or use a mechanical stirrer. Allow 30 seconds or so for the needle to stop drifting and read while stirring.

This sample may be saved for determination of the lime requirement.

Lime Requirement of Soils:

Reagents

1. *Buffer, pH 8.00*
   a. Dissolve 40 gm p-nitrophenol in approximately 350 ml hot distilled water.
   b. Dissolve 30 gm boric acid in approximately 350 ml hot distilled water.
   c. Dissolve 21 gm KOH pellets in approximately 25 ml distilled water.
   d. Dissolve 148 KCl in approximately 600 ml distilled water.
   e. To KCl solution add p-nitrophenol, boric acid, and KOH solutions, stirring well between each addition.
   f. Dilute to 2 liters.
   g. Adjust pH to exactly 8.00 with KOH or HCl as required.

Procedure

1. Add 20 ml buffer to soil sample on which pH was determined.
2. Stir thoroughly, let stand at least 10 minutes.
3. Set pH meter at 8.00 using 1:1 buffer-water solution.
4. Read pH while stirring sample. If pH is below 7.0 dilute with 20 ml buffer and 20 ml distilled water or repeat test with 10 gm soil sample; in either of these cases, the lime requirement on the following table must be doubled.
5. Lime requirements are given in Table 10.2.

Table 10.2: Lime Requirement to Bring Soil to pH 6.5 in Hundreds of Pounds Ag. Lime ($CaCO_3 \times 1.5$) per acre

Soil pH with Buffer	Soil pH															
	6.0	5.9	5.8	5.7	5.6	5.5	5.4	5.3	5.2	5.1	5.0	4.9	4.8	4.7	4.6	4.5
7.85	6	7	7	8	9	9	10	10	11	11	12	12	12	13	13	13
7.80	8	9	10	11	11	12	13	13	14	15	15	16	16	17	17	18
7.75	10	11	12	13	14	15	16	17	18	18	19	20	20	21	21	22
7.70	11	13	15	16	17	18	19	20	21	22	23	24	24	25	26	27
7.65	13	15	17	18	20	21	22	23	25	26	27	27	28	29	30	31
7.60	15	17	19	21	23	24	26	27	28	29	30	31	32	33	34	36
7.55	17	50	22	24	26	27	29	30	32	33	34	35	36	38	39	40
7.50	19	22	24	26	28	30	32	34	35	37	38	39	41	42	43	45
7.45	21	24	27	29	31	33	35	37	38	40	41	43	45	46	47	49
7.40	23	26	29	32	34	36	38	40	42	44	46	47	49	50	52	53
7.35	25	28	31	34	37	39	41	44	45	48	51	51	53	45	56	58
7.30	27	32	34	37	40	42	45	47	49	51	53	55	57	59	61	62
7.25	29	33	36	40	43	45	49	50	53	55	57	59	61	53	65	67
7.20	31	35	39	42	45	48	51	54	56	58	61	63	65	67	69	71
7.15	33	37	41	45	48	51	54	57	60	62	64	67	69	71	73	76
7.10	34	39	44	48	51	54	58	60	63	66	68	71	73	75	78	80
7.05	36	41	46	50	54	58	61	64	67	69	72	75	77	79	82	85
7.00	38	44	48	53	57	61	64	67	70	73	76	78	81	84	86	89

For example, if the soil pH is 5.2 and the pH with the buffer is 7.5, then 1587.6 kg of agricultural lime is required for each acre of pond bottom to bring soil pH to 6.5.

If twice the amount of buffer, or half the amount of soil was used then if soil pH were 5.2 and the pH after adding the buffer was 7.5, then 3175.2 kg agricultural lime is required per acre of soil.

To convert pounds/acre to kg/ha, multiply by 1.1. For example,

3500 pounds per acre × 1.1 = 3850 kg/ha.

Determination of Calcium in Pond Bottom Soils

Reagents

1. *Extracting solution*: Ammonium acetate – pH 7.00. To 500 ml of distilled water add 70 ml concentrated ammonium hydroxide and 60 ml concentrated acetic acid. Then add sufficient distilled water to make 1 liter. Let stand overnight and determine the pH. Adjust pH to 7.00 by adding NH_4OH or acetic acid as required.
2. *Aqua-Regia*: Mix 3 parts conc. HCl and part conc. HNO_3. Use immediately.
3. *Hydrogen peroxide, reagent grade (30 per cent).*
4. *HCl, concentrated.*
5. *NaOH Solution* (page 302).
6. *Murexide Indicator* (page 302).
7. *EDTA, standard* (page 301).

Procedures

A. *Preparation of the Soil Sample*

1. Collect samples of bottom soils from various parts of the pond and combine to form a well-mixed composite sample.
2. Remove a half-part sample and dry in an oven at 80°C.
3. Grind sample in mortar or in a mill until it passes a 20-mesh screen.

B. *Procedure for Extraction of Ca from Soils*

1. Weigh 5.0 gm of soil into a 250 ml beaker.
2. Add 20 ml ammonium acetate extracting solution. Stir for 5 minutes.

3. Filter through a Buchner funnel using asbestos fiber. Leach further with three 10 ml portions of extracting solution.
4. Place the ammonium acetate extract in a 250 ml beaker and heat on a hot plate or steam bath under a funnel hood. Avoid boiling during evaporating. Gently heat as solution nears dryness to avoid splattering of salts. Continue heating until dense fumes of ammonium acetate have been removed.
5. Remove beakers from hot plate and add 2 to 3 ml aqua regia to dissolve salts.
6. Reheat just to dryness to remove HCl.
7. Allow beaker to cool.
8. Add 2 ml of 30 per cent analytical grade hydrogen peroxide.
9. Warm but avoid excessive frothing. Evaporate slowly just to dryness. Do not overheat.
10. If organic matter persists after peroxide treatment (denoted by yellow colour) repeat peroxide treatment as necessary.
11. Add just enough concentrate HCl to bring all calcium salts into solution. Two ml HCl is usually enough. Add a little distilled water.
12. Transfer to a 100 ml volumetric flask and make up to 100 ml with distilled water.

C. *Procedure for Analysis:*

1. Place a 5 ml aliquot portion of the 100 ml total in a 250 ml Erlenmeyer flask.
2. Add 45 ml distilled water.
3. Add 2 ml of approximately NaOH solution and a pinch (approximately 0.2 ml) of Murexide.
4. Titrate with standard EDTA solution.
5. Towards the end point, stop, stir or shake for about 1 minute before adding the last few drops that will determine the end point.
6. One ml of EDTA is equivalent to 1 mg $CaCO_3$. Hence, total ml of EDTA used × 20 × 200 = ppm $CaCO_3$ in soil sample.

Determination of Total Hardness of Ammonium Acetate Extract of Soils

Procedure for Analysis

1. Place a 5 ml aliquot portion of the 100 ml total obtained in step 12 for Ca extraction from soils in a 250 ml Erlenmeyer flask.
2. Add 45 ml distilled water.
3. Add 2 ml NH_4Cl–NH_4OH buffer.
4. Add rapidly the same amount of standard EDTA that was used for Ca analysis of this sample (page 329). Stir or shake for a short while.
5. Add 2 to 3 drops of Eriochrome black T or Calmagite (page 301).
6. Titrate to the end point.
7. One ml of EDTA is equivalent to 1 mg of $CaCO_3$. Hence, total ml of EDTA used × 20 × 200 = ppm total hardness as $CaCO_3$.

Total Phosphorus (Estimated) in Soils

Acid – NH_4F – Soluble Phosphorus

Acid–NH_4F–soluble phosphorus included all calcium phosphates, total absorbed phosphorus, and almost all the phosphorus held as Fe and Al complexes. Hence, subtraction of dilute-acid-soluble phosphorus and neutral–NH_4F–soluble phosphorus from acid–NH_4F–soluble phosphorus gives a difference which represents phosphorus held as Fe and Al complexes.

Reagents

1. *0.1N HCl*: Dilute 17 ml conc. HCl to 2 liters with distilled water (approximate normality only).
2. *Ammonium Fluoride (NH_4F)*: Reagent grade.
3. *Ammonium molybdate-sulphuric acid solution*: Same as on page 297.
4. *Stannous chloride solution*: Same as on page 297.
5. *Boric acid*: Make a saturated solution of boric acid by dissolving approximately 50 gm boric acid in 1 liter distilled water.

Procedure

1. Shake 2 gm of soil sample with 100 ml 0.1N HCl for 30 minutes.
2. Add 2 gm of solid NH_4F. Weigh out this substance rapidly as it is hygroscopic.
3. Shake for one hour, using mechanical shaker.
4. Filter through No. 5 Whatman filter paper.
5. To a 10 ml aliquot, add 15 ml boric acid and dilute to exactly 50 ml. Add 2 ml ammonium molybdate-sulphuric acid solution and 5 drops stannous chloride. If phosphorus concentration is high, use a 5 ml aliquot instead of 10 ml and 15 ml of distilled water instead of 10 ml. Mix well. Let stand for 10 minutes.
6. Read per cent transmittance or optical density with a spectrophotometer at 690 millimicrons, and compare with calibration curve prepared as on pages 298-299.
7. Multiply ppm × 250 to obtain ppm in soil.

Adsorbed Phosphorus (Estimated) or Neutral-HN_4F-Soluble Phosphorus

Reagents

1. *Extracting solution*: Dissolve 18.5 gm ammonium fluoride in 1 liter of water. Adjust the pH to 7.0 with ammonium hydroxide solution.
2. *Ammonium molybdate-sulphuric acid solution*: Same as on page 297.
3. *Stannous chloride solution*: Same as on page 297.
4. *Boric acid*: Make saturated solution of boric acid by dissolving approximately 50 gm boric acid in 1 liter of water. (same as on page 331).

Procedure

1. Shake 2 gm of soil sample with 100 ml of extracting solution for 1 hour.
2. Filter through No. 5 Whatman filter paper.
3. To a 20 ml aliquot, add 15 ml boric acid, and dilute to exactly 50 ml. Then add 2 ml molybdate solution, and 5

drops stannous chloride solution, in this order. If soil is high in phosphorus, use a 10 ml aliquot. Mix well. Let stand for 10 minutes.

4. Read per cent transmittance or optical density by means of a spectrophotometer at 690 millimicrons, and compare with the calibration curve as prepared on pages 298-299.
5. Multiply ppm × 125 to obtain ppm in soil.

Dilute – Acid – Soluble Phosphorus (Available Phosphorus) in Soils

This inorganic phosphorus fraction of soils represents most of the phosphorus present as calcium phosphate.

Reagents

1. *Approximately 0.1 N H_2SO_4*: Prepare approximately 1N H_2SO_4 as described under Reagent 1, page 334. Pipette 100 ml of this reagent in to a liter flask and dilute to the mark with distilled water.
2. *Extracting solution*: In a 1 liter flask dissolve 3 gm of ammonium sulphate in 900 of distilled water, add 20 ml of 0.1.N sulphuric acid, and dilute to 1 liter with distilled water.
3. *Molybdate solution*: Ammonium molybdate-sulphuric acid as prepared on page 297.
4. *Stannous chloride solution:* Prepared as on page 297.

Procedure

1. Shake 2 gm of soil sample with 200 ml of extracting solution for 30 minutes.
2. Filter through No. 5 Whatman filter paper.
3. Dilute a 40 ml aliquot to exactly 50 ml, add 2 ml molybdate solution and then 5 drops stannous chloride. Mix well. Let stand for 10 minutes.
4. Read per cent transmittance or optical density with a spectrophotometer at 690 millimicrons and compare with calibration chart prepared as on page 298-299.
5. Multiply ppm × 125 to obtain ppm in soil.

Analysis of Pond Muds

Ammonia Nitrogen

Reagents

1. *Approximately 1 N H_2SO_4**: To 800 ml distilled water in a 1 liter volumetric flask, add 27.8 ml conc. H_2SO_4 slowly, allowing it to flow down inside the flask. Mix and dilute to 1 liter.
2. *Potassium sulphate-sulphuric acid solution*: Dissolve 174.27 gm of pure K_2SO_4 in distilled water. Add 20 ml of approximately 1 N H_2SO_4 and dilute to 2 liters. This solution contains 1 N K_2SO_4 and 0.01N H_2SO_4.
3. *Sodium hydroxide*: Dissolve 400 gm of pure NaOH in distilled water and dilute to 1 liter. This is approximately a 10 N solution.
4. *Hydrochloric acid*: Dilute 17 ml concentrated HCl to 2 liters with distilled water. This solution is approximately 0.1 N.
5. *Antifoam*: Add 1 part paraffin oil to 1 part octly-alcohol.
6. *Nessler's reagent*: This reagent can be purchased or made as described on page 286.
7. *Standard NH_4 solution*: Make as directed on page 286.

(* Concentrated H_2SO_4 is approximately 36 N, with a specific gravity of 1.84.)

Procedure

1. Weigh 10 gm of air-dried soil and put into a 100 ml beaker.
2. Add 25 ml of the $K_2SO_4 - H_2SO_4$ solution, stir to insure intimate contact, and let stand for 2 hours.
3. Transfer the sample to a Buchner funnel, using a mesh 5 filter, and filter into a 250 ml Erlenmeyer flask. This procedure requires the use of vacuum apparatus.
4. Leach the sample with 125 ml addition a $K_2SO_4 - H_2SO_4$ solution, allowing the soil to drain between each addition. (Keep soil for Organic N Analysis).
5. Dilute the filtrate to 250 ml and mix.
6. Transfer a 100 ml aliquot into a 500 ml Kjeldahl flask : the blank flask will contain 100 ml of the K_2SO_4-H_2SO_4 solution.

7. Add 100 ml distilled water, boiling chips[#], drops of antifoam, and 10 ml of NaOH.
8. Distill into a 250 ml Erlenmeyer flask containing 10 ml of HCl until about 150 ml of distillate is collected. (keep residue for $NH_3 + NO_2$ analysis).
9. Dilute the distillate to 500 ml with distilled water.
10. Nesslerize a 50 ml portion and evaluate ppm by spectrophotometer (page 295).
11. Multiply ppm measured by 125 to obtain ppm in soil sample.

    ([#]Carborundum granules, teflon chips or short sections cut from glass tubing can be used to prevent 'bumping'.)

Nitrate + Nitrite Nitrogen

Reagents

1. *Anti-foam*: Add 1 part paraffin oil to 1 part octyl-alcohol.
2. *Devarda's alloy metal*: 49 per cent copper, 45.5 per cent aluminum, and 4.7 per cent zinc.
3. *Hydrochloric acia*: Dilute 17 ml concentrated HCl to 2 liters with distilled water. This solution is approximately 0.1N.
4. *Nessler's reagent*: This reagent can be purchased or prepared as on page 286.

Procedure

1. Add 200 ml distilled water to the residue in the Kjeldahl flaks (samples blank) left over from the NH_3 analysis.
2. Add 1 drop of anti-foam and 0.4 gm powdered Devarda's alloy metal.
3. Distill sample and blank into a 250 ml Erlenmeyer flask containing 10 ml HCl until about 125 ml of distillate is collected.
4. Dilute the distillate to 200 ml.
5. Nesslerize a 50 ml portion and evaluate by spectrophotometer (page 293)
6. Multiply ppm by 50 to obtain ppm in sample.

Organic Nitrogen in Pond Muds

Reagents

1. *Sulphuric acid*: Concentrated H_2SO_4.
2. *Sodium sulphate-mercuric oxide reagent*: Mix 454 gm of Na_2SO_4 with 24.4 gm of HgO.
3. *Sodium hydroxide*: Prepare as described on page 334.
4. *Zinc metal*: Granulated zinc.
5. *Hydrochloric acid*: Dilute 17 ml concentrated HCl to 2 liters with distilled water. This solution is 0.1 N.
6. *Nessler's Reagent*: The reagent can be purchased or made as described on page 286.

Procedure

1. Transfer leached soils and filter papers (page 334) into 800 ml kjeldahl flask; the blank contains 1 No. 5 filter paper.
2. Add 40 ml H_2SO_4 and a teaspoon of Na_2SO_4-HgO reagent.
3. Digest until the residue is white (blank is clear) and an additional 10 minutes.
4. Cool, add 200 ml distilled water, mix and cool again and 150 ml of NaOH.
5. Add a pinch of zinc granules and 150 ml of NaOH.
6. Distill into a 250 ml Erlenmeyer flask containing 20 ml of HCl, until about 150 ml of distillate is collected.
7. Dilute the distillate to 500 ml.
8. Dilute 20 ml of the distillate to 1,000 ml.
9. Nesslerize a 50 ml portion and evaluate by spectrophotometer (page 293).
10. Multiply measured ppm by 2,500 to obtain ppm in the original soil.

    10 g soil = Distillate 500 ml (500 grams), step 7 above is a dilution of 1 to 50.

    Step 8, 20 ml to 1000 ml is a second dilution of 1 to 50.

    Total dilution = 50 × 50 = 2500.

Appendices

Selected Atomic Weights

A	39.94	I	126.92
Ag	107.88	K	39.10
Al	26.97	Mg	24.32
As	74.91	Mn	54.93
Au	197.20	N	14.01
B	10.82	Na	23.00
Ba	137.36	O	16.00
Br	79.92	P	31.02
C	12.00	Pb	207.22
Ca	40.08	S	32.06
Cl	35.46	Se	78.96
Cu	63.57	Si	28.06
F	19.00	Sn	118.70
Fe	55.84	Sr	87.63
H	1.008	U	238.14
He	4.00	Zn	65.38
Hg	200.61		

Conversion
Centigrade to Fahrenheit Degree

°C	°F	°C	°F	°C	°F
-18	-0.4	8	46.4	29	84.2
-16	+3.2	9	48.2	30	86.0
-14	+6.8	10	50.0	31	87.8
-12	+10.4	11	51.8	32	89.6
-10	+14.0	12	53.6	33	91.4
-8	+17.6	13	55.4	34	93.2
-7	+19.4	14	57.2	35	95.0
-6	+21.2	15	59.0	36	96.8
-5	+23.0	16	60.8	37	98.6
-4	+24.8	17	62.6	38	100.4
-3	+26.6	18	64.4	39	102.2
-2	+28.4	19	66.2	40	104.0
-1	+30.2	20	68.0	41	105.8
0	+32.0	21	69.8	42	107.6
1	33.8	22	71.6	43	109.4
2	35.6	23	73.4	44	111.2
3	37.4	24	75.2	45	113.0
4	39.2	25	77.0	46	114.8
5	41.0	26	78.8	47	116.6
6	42.8	27	80.6	48	118.4
7	44.6	28	82.4	49	120.2
				50	122.0

Conversion Grams to Pounds

Grams	*Pounds*	*Grams*	*Pounds*
1	0.0022	120	0.2645
2	0.0044	140	0.3086
3	0.0066	160	0.3526
4	0.0088	180	0.39.67
5	0.0110	200	0.4408
6	0.0132	220	0.4849
7	0.0154	240	0.5290
8	0.0176	260	0.5730
9	0.0198	280	0.6171
10	0.0220	300	0.6612
20	0.0441	320	0.7053
30	0.0661	340	0.7494
40	0.0882	360	0.79.34
50	0.1102	380	0.83.75
60	0.1322	400	0.8816
70	0.1543	420	0.9257
80	0.1763	440	0.9698
90	0.1984	453.6	1.0000
100	0.2204		

Relation Between Transmittancy (TS) and Optical Density (D)

T_s (%)	D	T_s (%)	D	T_s (%)	D	T_s (%)	D
100	0.000	75	0.125	50	0.301	25	0.602
99	0.004	74	0.131	49	0.310	24	0.620
98	0.009	73	0.137	48	0.319	23	0.638
97	0.013	72	0.143	47	0.328	22	0.658
96	0.018	71	0.149	46	0.337	21	0.678
95	0.022	70	0155	45	0347	20	0.699
94	0.028	69	0.161	44	0.357	19	0.721
93	0.032	68	0.168	43	0.367	18	0.745
92	0.036	67	0.174	42	0.377	17	0.770
91	0.041	66	0.181	41	0.387	16	0.796
90	0.46	65	0.187	40	0.398	15	0.824
89	0.051	64	0.194	39	0.409	14	0.854
88	0.056	63	0.201	38	0.420	13	0.886
87	0.061	62	0.208	37	0.432	12	0.921
86	0.066	61	0.215	36	0.444	11	0.959
85	0.071	60	0.222	35	0.456	10	1.000
84	0.076	59	0.229	34	0.469	9	1046
83	0.081	58	0.237	33	0.482	8	1.097
82	0.086	57	0.244	32	0.495	7	1.155
81	0.092	56	0.252	31	0.509	6	1.222
80	0.097	55	0.260	30	0.523	5	1.301
79	0.102	54	0.268	29	0.538	4	1398
78	0.108	53	0.276	28	0.552	3	1.523
77	0.114	52	0.284	27	0.569	2	1.699
76	0.119	51	0.292	26	0.585	1	2.000

Standard and Conversion Factors

Metric Prefixes

Nano	10^{-9}	used with meter* (length)
Micro	10^{-6}	used with gram (weight)
Milli	10^{-3}	used with liter (volume), gram (weight)
Centi	10^{-2}	
Deci	10^{-1}	
Daka	10	
Hecto	10^{2}	
Kilo	10^{3}	
Mega	10^{6} (rarely used)	

* micrometer = micron

nanometer = millimicron.

Length

Inch	2.54 centimeters = 25.4 millimeters
Foot	30.48 centimeters = 0.305 meter
Yard	3 feet = 0.914 meter
Mile	5,280 feet = 1,760 yards = 1,609 kilometers = 1,609.35 meters
Fathom	6 feet = 1.829 meters
Knot	1.15 miles per hour
Micron	0.001 millimeter
Centimeter	10 millimeters = 0.39 inch = 10,000 microns
Meter	3.28 feet = 39.67 inchs = 1.094 yards = 0.546 fathoms
Kilometer	3,280.83 feet = 0.621 mile

Area

Square inch	6.451 square centimeters = 0.0069 square feet
Square foot	144 square inches = 0.0929 square meters
Square yards	9 square feet = 0.836 square meter
Acre	43,560 square feet = 4,840 square yards = 4,046.87 square meters = 0.405 hectare
Section	1 square mile = 640 acres = 259 ha.
Square mile	640 acres = 2.59 square kilometers = 259 ha.
Square centimeter	0.155 square inch

Square meter	10,000 square centimeters = 10.76 square feet = 2.47 × 10^{-4} acres
Square kilometer	247.1 acres = 1.076 × 10^7 square feet = 0.386 square miles
Hectare	10,000 square meters = 2.47 acres = 100 acres
Acre	0.01 hectare = 0.025 acre

Volume

Cubic inch	5.787 × 10^{-4} cubic feet = 16.38 ml = 0.004 gallon = 0.016 liter
Cubic foot	0.037 cubic yard = 0.028 cubic meters = 7.48 gallon = 28.32 liters
Cubic yard	27 cubic feet = 0.765 cubic meters
Cubic mile	4.165 cubic kilometers
Acre-foot	43,560 cubic feet = 1,219.68 cubic meters = 325,828.8 gallons
Minim	0.06 ml = 1/60 dram
Dram (fluid)	3.69 ml = 0.225 cubic inch = 60 minims
Teaspoon	0.33 tablespoon = 4.9 ml
Tablespoon	3 teaspoon = 14.7 ml
Ounce (fluid)	1.804 cubic inch = 29.57ml = 0.029 liter
Cup	14.4 cubic inches = 236 ml
Pint (liquid)	28.88 cubic inches = 473.18 ml = 2 cups = 0.473 liter
Quart (liquid)	57.75 cubic inches = 0.03 cubic feet = 0.946 liter = 2 pints
Gallon	231 cubic inches = 0.134 cubic feet = 3.785 liters = 4 quarts
Bushel	1.24 cubic feet = 2,150 cubic inches = 35.24 liters
Cubic centimeter	0.061 cubic inches = 0.9997 ml
Cubic meter	35.31 cubic feet = 1.31 cubic yards = 264.17 gallons = 999.97 liters
Cubic Kilometer	1 × 10^9 cubic meters = 0.239 cubic mile
Liter	61.03 cubic inches = 0.035 cubic foot = 0.264 gallon = 33.81 ounces (fluid)

Weight (Avoirdupois)

Grain	0.065 gm
Dram	27.3 grains = 1.77 gm
Ounce	16 drams = 437.5 grains = 28.35 gm = 0.0625 pound
Pound	16 ounces = 7,000 grains = 453.59 gm = 0.454 kilogram

Ton (short)	2,000 pounds = 0.907 metric ton = 907.18 kilograms
Ton (long)	2,240 pounds = 1,016.05 kilograms

Pounds per acre × 1.12 = kilograms per hectare

Gram	0.56 dram–0.035 ounce–0.0022 pound
Kilogram	35.27 ounces = 2.204 pounds
Ton (metric)	2,204 pound = 1,000 kilograms

Water

Boiling point	100° C, 212° F at 760 mm air pressure
Freezing point	0° C, 32° F

1 Atmosphere = 33.8995 feet of water @ 4° C = 14.696 pounds square inch.

Water pressure : (Depends upon vertical height of water column)

1 inch height @ 39.2° F (4° C) = 0.578 ounces/ square inch = 5.20 pounds/ square foot

1 foot height @ 39.2° F (4° C) = 0.433 pounds/ square inch = 62.427 pounds/ square foot

Weight of water : (Maximum density at 4° C or 39.2° F)

1ml	1 gram @ 4° C; 0.99987 @ 0° C; 0.958 @ 100° C
1 cubic foot	62.427 pounds @ 4° C; 62.24 pounds @ 25° C
1 acre-foot	2.71×10^6 pounds (@ approximately 25° C–30° C)
1 cubic mile	3.3792×10^6 acre-feet = 9.157×10^{12} pounds = $4.57^9 \times 10^9$ tons
1 pint	1.04 pounds
1 quart	2.08 pounds
1 gallon	8.34 pounds
1 cubic meter	2,203.3 pounds

Density of Gases

O_2 at STP (0° C, 760 mm Hg) 1.429 gm/l

CO_2 at STP (0° C, 760 mm Hg) 1.9369 gm/l

To convert from ml gas per liter water to ppm multiply by the density of the gas at standard temperature and pressure (STP). *e.g.* 10 ml O_2 /1 × 1.429 = 14.29 ppm O_2

Alkalinity

To convert other units to ppm $CaCO_3$ or ppm CO_3^{--} multiply by the following factor :

Unit	*ppm $CaCO_3$*	*ppm CO_3^{--}*
Grains $CaCO_3$/ U.S. gallon	17.1	10.25
Grains $CaCO_3$/ English gallon	14.3	8.58
ppm CaO	1.79	1.07
ml 0.1 N HCl/100ml sample	50	30
milliequivalent/ liter	50	30
equivalent ppm	50	30
ppm $CaCO_3$	1	0.60
ppm CO_3^{--}	1.67	1

Glossary

Acclimatization: Insuring to a new habitat, medium or set of conditions.

Acid: Having a somewhat sour quality.

Air-breather: A fish capable of directly utilizing atmospheric oxygen for respiratory purposes.

Algae: Simple chlorophyll-bearing plants most of which live entirely submerged in water, fresh or salt.

Algal bloom: A preponderance of algae in water which is turned completely green.

Algal scum: Floating surface layer of algae in the fresh or decaying condition.

Alkaline: Having an acrid taste and soapy touch.

Alkalinity: A measure of the dissolved lime (calcium) content of water.

Anal fin: The unpaired fin that lies along the middle line on the ventral surface of the body behind the anus or vent.

Anchorage: A basis for attachment.

Animalcule: A minute animal which cannot ordinarily be seen with the unaided eye.

Anterior: Towards the front or head end.

Aquaria: Containers with glass frontage and/or sides in which aquatic organisms could be observed and reared.

Artificial food: Food not produced in the natural habitat; specially introduced food.

Asphyxiation: Suffocation; death due to respiratory difficulties; *i. e.* inadequate exchange of oxygen and carbon dioxide.

Bacteria: Extremely small, single-celled fungoid plants, reproducing rapidly, associated with decomposition of substances and regarded as germs of diseases.

Barbell: A soft, slender, fleshy, whisker-like feeler near the mouth in some fishes; a fish belonging to the genus *Barbus*.

Bati: A small cup-like measure usually of 130 to 150 cubic centimeter capacity, used for measuring early carp fry.

Bhasa-badha fishery: Low-lying, marshy, salt water areas under reclamation in the Sunderbans, subject to tidal-effect; embanked and provided with sluices devised to trap in fish-seed to grow.

Biophysical: Pertaining to the living organisms and water conditions of the environment.

Bloom: Vide Algal bloom. A sudden appearance of minute algae in large numbers in waters.

Bottom fauna: Animal organisms usually living at the bottom of ponds and other waters.

Bottom feeder: A fish that generally feeds at the lower levels of water immediately above the bottom.

Brackish: Water which is intermediate in taste between salt and fresh water.

Breeder: A sexually mature fish ready to breed.

Breeding pond: Special pond in which fish are introduced for breeding.

Brood: Group of young fish of the same age and parentage.

Bundh: Irrigation tank of certain south western districts of W. Bengal in which the major Indian carps breed during the rainy months.

Cannibalistic: Tendency of eating one's own kind.

Carbon dioxide: A gas formed of carbon and oxygen and evolved by animal respiration and combustion.

Carnivorous: Eating animal food, living or dead.

Carp: A group of fresh water fishes with toothless mouth.

Catfish: One of a group of scaleless fishes, often with long whisker-like feelers.

Caudal fin: Tail fin.

Caudal peduncle: The hind region of the body from the anal fin to the base of the tail fin.

Caudal spot: Pigment spot at the base of the tail fin or at the tip of the caudal peduncle.

Chironomid: An insect of the order Diptera whose young ones, blood-red and worm-like, are invariably found at the bottom of ponds living in tubes.

Chlorinated: Treated with chlorine.

Chromatophore: One of the pigment cells (minute spots) usually found on the skin (of fish).

Climbing perch: (*Anabas testudineus*) So called because of its occasional habit of climbing trees and seeking similar situation.

Surface column feeder: A fish that generally feeds at the surface or in mid-deep water.

Compatible species: Fishes which can live and grow together in a pond with praying upon one another or competing among themselves for their food.

Conditioning: Processing of keeping young fish in a limited space for some time, to get the gut voided or contents and also to accustom them to the limited space.

Cormorant: A species of fish-eating water bird which dives in to catch fish.

Crustacea: A group of animals with jointed limbs and body and hard external shell; *e.g.* prawns, shrimps, and crabs.

Cultivable: That which can be cultivated; that in which fish could be grown under controlled conditions.

Culture: Process of cultivation of fish in all its stages.

Cyst: A small bag-like structure containing quiescent or development stages of parasites.

Derris powder: A commercial insecticide made of the powdered roots of the plant *Derris elliptica*, used as a fish poison also.

Diet: A specific type or combination of prescribed food.

Doctor Fish: The tench, *Tinca tinca*, so called as its body slime is believed to have a curative effect on wounds.

Dorsal: Pertaining to the back or upper part of the body.

Dorsal fin: The unpaired fin along the middle line of the back.

Dragon-fly: A predatory flying insect (order Odonata) which is aquatic in habit while young and immature.

Dragonfly-nymph: The young aquatic crawling stage of the dragon-fly.

Dry bundh: A bundh tank which dries up during summer.

Duck-weeds: Group of floating plants belonging to the genera *Lemna, Azolla, Wolffia*, etc.

Ecological: Pertaining to the environment or habitat of animals and plants.

Egg membrane: The thin outer envelope of the egg which swells up after fertilization to form the vitelline membrane.

Embryo: The young one (fish) developing within the egg membrane.

Empirical: Elementary, based on trial and error methods.

Emulsion: A milky liquid prepared by mixed oil and water by means of (here) soap.

Equilibrium: Balanced or normal position.

Exotic species: Species of fishes not native to the country; introduced from other countries.

Fecundation: The process of impregnation of fertilization of the egg.

Fecundity: Prolificness of progeny or the large production of eggs by the female (fish) .

Fertilization: Same as fecundation.

Filamentous: Slender or thread-like.

Fimbriated: Fringed or with short fimbriae or processes.

Fin: The organ by which the fish balances itself and swims.

Fin rays: The segmented, usually flexible, branched or unbranced rods supporting the fin membrane.

Fin rot: A disease affecting usually the fins of fishes.

Fingerling: Young fish, one inch to six inches long.

Fishery: The business of/or place for catching fish.

Fish seed: Fertilized eggs, fry or fingerlings of fish used for growing in fish culture.

Forage fish: Small-sized fish generally utilized as food by larger predatory fishes.

Formalin: A solution of commercial formaldehyde used as preservative of animal and plant tissues.

Fry: Young fish up to one inch long.

Fungus: One of the lowly groups of plants lacking the green colouring matter of higher plants (Mushrooms, toadstools and moulds.

Gamcha: A piece of cloth used as tail cloth for the fry catching net, or for netting fry and fingerlings.

Gas disease: A fish disease caused by excess of nitrogen or oxygen, resulting in accumulation of gas in body spaces.

Gasteropod (Gastropod): A class of usually spiral-shelled molluscs with a flat creeping foot.

Genus: A group consisting of a number of closely related species.

Gill: Organ by which fish and other animals breathe in water.

Gill membrane: The thin layer covering the gills and permitting exchange of gases for respiratory purposes.

Gil rakers: Stiff long appendages developed on the inner margin of the hoop-like gill arch.

Gillrot: A disease commonly affecting the gills.

Gregarious: Living together or swimming about in shoals.

Gull: A fish eating bird of the sea coast and river estuaries.

Git: The alimentary canal through which food passes in digestion, assimilation and excretion.

Hair weed: A genus of long and filamentous green algae, viz., *Spirogyra.*

Hapa: A miniature tank-like structure made of cheap cloth and fixed to poles in water for temporary storage of fish seed.

Hatching: The process of bursting of the egg membrane and liberation of the baby fish.

Hatchling: Young one just hatched out of the egg.

Herbivorous: Eating or living on plants, gross or minute.

Hexagonal: Six-sided.

Hundi: A wide mouthed vessel made of mud or tin used for transporting fish seed.

Indigenous: Occurring naturally in a country ; native to.

Inlet: Passage by which water is allowed to flow in.

Inorganic: Without life ; of mineral origin.

Insecticide: An insect-killer made of plants or chemicals, of organic or inorganic origin.

Insectivorous: Live on insects.

Iris: The coloured part of the eye.

Iridescence: Exhibition of colours like those of a rainbow.

Larva: Young one (of fish) after hatching, till the yolk is fully absorbed.

Lateral line: A longitudinal series of pores along each side of the body of the fish appearing as a dotted line.

Leather Carp: A varietyof the common carp (*Cyprinus carpio*) characterized by the almost complete absence of scales.

Leptocephalus: Flat, transparent, ribbon-shaped young one of an eel.

Lethal: Deadly.

Life-cycle: The stages in the life of an organism from birth to death.

Live fish: A fish that is capable of living outside water for some time by virtue of its capacity to breathe atmospheric air.

Long-seasonal pond or tank: A sheet of water that available for fish cultivation during six to eight months in a year.

Macro-vegetation: Large-sized plants, as opposed to micro-vegetation or minute plants

Major Carps: One of the larger carps attaining a length of over two feet.

Microscopic: Minute, usually visible only under a microscope.

Minnow: A fresh-water fish which does not generally attain a size larger than that of fry or fingerling of larger carps.

Minor carp: One of the smaller carps as opposed to the large-sized major carp.

Mirror Carp: A common variety of the Common carp, *Cyprinus, carpio*, having large shining mirror-like scales on the body.

Mixed farming: The practice of cultivation more than one species of compatible fishes in the same pond at the same time.

Molluscs: A soft-bodies and shelled animal, e.g., the common pond snail, or mussel belonging to one of the divisions of Invertebrate animals.

Mucilaginous: Slimy.

Mulletry: A fish farm where mullets form the important crop raised.

Murrel: One of the common live fishes with the head remotely resembling that of a snake and hence called snake-headed (*Channa* spp.).

Nape: The back or upper part of the neck.

Nitrate: One of the inorganic fertilizers.

Nitrogen: A gas farming nearly four-fifths of the atmosphere.

Non-predator: One (Fish) that does not kill others for food.

Nusery pond: A pond specially prepared for rearing baby carps from the early stage to advanced fry or early fingerlings stage.

Ocellus: A small round eye-like spot on the fin.

Occipital: Pertaining to the occiput-the hindmost region on the top of the head.

Omnivorous: Feeding on animal, vegetable or mineral food.

Oozing: The fully ripe conditions of the gonads in which the milt of the male as well as eggs of the female are ready to be shed.

Ostracod: A small group of minor crustaceans.

Otter: A common fish-eating carnivore of aquatic habits.

Outlet: Passage for water to flow out.

Ovary: The female reproductive organ in which egg cells develop.

Oxygen: An essential gas forming part of the atmosphere which supports life.

Parasite: An organism in or on the body of other organisms deriving its sustenance from its host.

Parasitization: The process of infection by a parasite.

Parental care: The practice of looking after the eggs and young ones exhibited by some fishes and other animals.

Pearl spot: The brackish water perch *Etroplus suratensis*.

Pectoral fins: The front or uppermost of the paired fins, at the level of the shoulder.

Pectoral spines: Modified, stronger rays of the pectoral fins.

Perch: A fish of a group, with strong spines on the dorsal and anal fins which are generally long; salt water as well as fresh water in habitat.

Perennial (pond): A pond the water in which does not dry up completely during summer.

Phosphate: A common inorganic manure.

Phytoplankton: Plant organisms of plankton.

Pisciculture: The process of cultivation of fish.

Piscivorous: That which eat fish as food.

Plankton: Population of minute plants and animals with limited movement which passively float in the surface layers of fresh or salt water.

Predator: That which kills other animals for food.

Prolific: With high reproductive capacity.

Protective food: Food which contains the essential requirements for body building.

Protein: An essential for of complex composition in which the flesh of fish is rich.

Rearing pond: Pond used for rearing advanced fry to fingerlings stage.

Rotenone: The active principle in the fish poison Derris powder.

Salinity: The total salt content of water.

Scale: One of the numerous, thin, plate-like structures covering the body of fish.

Scale carp: A variety of the common carp *Cyprinus carpio*, having scales covering the entire body surface.

Seasonal (pond): A pond which contains water only for a part of the year, usually between four to eight months.

Short-seasonal: A pond which will contain water only for three to four months in a year.

Shoulder spot: A pigment spot near the pectoral or shoulder region seen in some fishes.

Shrimps: A small prawn-like crustacean.

Snout: The part of the head in front of the eye.

Soundings: Depth records.

Spawn: To lay or produce eggs, especially of fishes; also the laid eggs of fishes and other lower animals.

Spawning ground: Selected places to which fishes migrate for spawning.

Species: A group of individual specimens having close resemblance but differing from others and belonging to the same genus.

Sporting fish: A fish that takes bait and can be caught on a hook.

Stagnating: Standing, not flowing.

Stock figure: The number of fish that should be stocked in a particular pond depending on its productivity.

Stocking pond: A pond for stocking fingerlings of one year old fishes.

Stripping: The process of pressing out ripe roe or milt of fishes for artificial fecundation.

Submerged: Sunk under water.

Sucker: An adhesive organ by which a parasite attaches itself to the body of fish.

Super-saturation: Over-abundance.

Surface feeder: A fish that generally feeds in the surface layers of water.

Table fish: Edible fish of appreciable size.

Tail cloth: The open trough-like cloth at the hind end of the fry catching net in which fry accumulate.

Thinning: Removing part of the fish stock from a pond.

Total length: The greatest straight line distance between the tip of the snout and the tip of the tail fin.

Tow-net: Conical net of fine cloth used for filtering plankton from water.

Toxic: Poisonous.

Toxicity: Poisonous property.

Understocking: Stocking below the capacity of the pond.

Unicellular: Single-celled.

Ventral: Pertaining to the lower surface of the body.

Ventral fins: The paired fins situated below and behind the pectrol fins; also called the pelvic fins.

Virgin water: Water not previously utilized for systematic fish culture.

Viscocity: The property of being thick and not flowing quickly.

Water-fleas: Micro-crustaceans largely used as food by baby fish.

Weed: Unwanted plants at the bottom sides and surface of the pond or bundh, or at or near the margin of water.

Weed fish: Minnows and other fishes which serve as food of other selected fishes in pisiculture.

Wet bundh: A perennial irrigation pond used for carp breeding.

Wier: A platform or side usually built of stone at the outflow end of an irrigation tanks or bundh to allow excess water to overflow after the highest level has been reached.

Yearling: A fish about one year old.

Yolk-sac: Sec containing the stored food of the egg on which the hatching subsists for a few days.

Bibliography

Alikunhi, K.H. 1956. Fish culture techniques in India, P. 63-73. In : *Progress of fisheries development in India*. Central Inland Fisheries Research Institute and Department of Fisheries, Orissa, India.

Appelbaum, S. and U. Dor. 1978. Ten-day experiment nursing of carp (*Cyprinus carpio* L.) larvae with dry feed. *Bamidgeh* 30(3) : 85-88.

Appelbaum, S. and B. Uland. 1979. Intensive rearing of grass crap larvae *Ctenopharyngodon idella* (Valenciennes 1844) under controlled conditions. *Aquaculture* 17 : 175-179.

Bhowmick, R.M. 1978. Glassjar hatchery for Indian major carps and Chinese carps. *Indian J. Fish.* 25(1/2) : 171-179.

Bryant, P.L. and A.J. Matty. 1980. Optimisation of Artemia feeding rate for carp larvae (*Cyprinus carpio* L.). *Aquaculture* 21 :203-212.

Chang, W.Y.B., J.S. Diana and W. Chuapoehutz. 1983. *Workshop report to agency for international development*, 19-29 April 1983. Strengthening of South-east Asian aquaculture Institutions (Grant No. DAN-5543-G-SS-2103-00). 30 p. Mimeo.

Chow, K.W. 1980, Microencapsulated egg diets for fish larvae, P. 355-361. In : *Fish feed technology*. ADCP/REP/8011. Aquaculture Development and Coordination Programme, FAO, Rome.

Gohl, B. 1975, *Tropical feeds*, FAO, Rome.

Harvey, B.J. and W.S. Hoar. 1979. *Theory and practice of induced breeding in fish.* International Development Research Centre, Ottawa, Canada.

Hora, S.L. and T.V.R. Pillay. 1962. *Handbook on fish culture in the Indo-Pacific Region.* FAO Fish. Tech. Pap. 14. 204 p.

Horwitz, W. 1980. *Offical methods of analysis of the Association of Official Analytical Chemists (A.O.A.C.).* A.O.A.C., Washington.

Houssay, B.A. 1931. Action sexuelle de l'hypophyse sur les poisons et les reptiles. C.R. *Soc. Biol.* 106: 377-378.

Jauncey, K. 1982. Carp (*Cyprinus carpio* L.) nutrition : A review. P. 216-263. In : J.F. Muir and R.J. Roberts (eds.) *Recent advances in aquaculture.* Croom Helm, London.

Jauncey, K. and B. Ross. 1982. *A guide to tilapia feeds and feeding.* Institute of Aquaculture, University of Stirling, Stirling, Scotland.

Khan, H.A. and V.G. Jhingran. 1975. Synopsis of biological data on rohu, *Labeo rohita* (Hamilton, 1822). *FAO Fish. Synop.* 111. 100p.

Lubzens, E., G. Sagie, G. Minkoff. E. Meragelman and A. Schneller. 1984. Rotifers (*Brachionus plicatilis*) improve growth rate of carp (*Cyprinus carpio*) larvae. *Bamidgeh* 36(2) : 41-46.

Nehrin, D. 1965. Die Ansnutzung verschiedener Getreidearten und Hulsenfruchte durch Karpfen. *Z.F. Fish, 13NF* 181-190.

Nose, T. 1978. Summary report on the requirements essential amino acids for carp. *EIFAC Symp. Finfish Nutrition and Feed Technology, Hamburg,* June 1978. EIFAC/78/Symp. E/8. FAO, Rome.

NRC, (National Research Council, Subcommittee on Warmwater Fish Nutrition). 1977. *Nutrient requirements of warmwater fishes.* National Academy of Sciences, Washington. D.C.

Ogino, C. 1980. Requirement of carp and rainbow trout for essential amino acids. *Bull. Jap. Soc. Sci. Fish,* 46(2) :171-174.

Ramchandran, V. 1963. The method and technique of using anhydrous ammonia for aquatic weeds control. *Proc. Indo-Pac. Fish Counc.* 10: 146-153.

Rothbard, S. 1982. Induced reproduction in cultivated cyprinids-the common carp and the group of Chinese carps. II. The rearing of larvae and the primary nursing of fry. *Bamidgeh* 34(1): 20-32.

Schoonbee, H.J. and F. de W. Brandt. 1982. Observations on some techniques employed for the removal of egg adhesiveness of the common carp. *Water S.A.* 4 : 93-103.

Schoonbee, H.J. and J.F. Prinsloo. 1984. Techniques and hatchery procedures in induced spawning of the European common carp, *Cyprinus carpio* and the Chinese carps *Ctenopharyngodon idella*, *Hypophthalmichthys molitrix* and *Aristichthys nobilis* in Transkei, *Water. S.A.* 10; 36-39.

Snell, J.R., A.Q. Chowdhury, W.M. Bailey and K.E. Sneed. 1979. Design features of a carp hatchery in Bangladesh. In : L.J. Allen and E.C. Kinney (eds.) *Proceedings of the bio-engineering symposium for fish culture, Michigan*, 16-18 October 1979. American Fisheries Society, Bethesda, Maryland.

Subramanian, S. 1983. Eradication of fishes by application of ammonia. *Aquaculture* 35: 273-275.

Swingle, H.S. 1947. Experiment in pond fertilization. *Ala. Agric Exp. Stn. Bull.* 264. 34p.

Ufodike, E.B.C. and A.J. Matty. 1983. Growth responses and nutrient digestibility in mirror carp (*Cyprinus carpio.*) fed different levels of cassava and rice. *Aquaculture* 31:41-50.

Vanhaecke, P. and P. Sorgeloos. 1983. International study on Artemia. XXX. Bio-economic evaluation of the nutritional value for carp (*Cyprinus carpio* L.) larvae of nine Artemia strains *Aquacutlure* 32: 285-293.

Woynarovich, E. 1969. Techniques of hypophysation of common carp, p. 1-11. In : FAO/UNDP Regional seminar on induced breeding of cultivated fishes.

Woynarovich, E. 1975. *Elementary guide to fish culture in Nepal.* FAO, Rome.

Woynarovich, E. and A. Woynarovich. 1980. A modified technology for elimination of stickiness of common carp (*Cyprinus carpio*) eggs. *Aquacult. Hung.* 2: 19-21.

Woynarovich, E. and L. Horvath. 1980. Artificial propagation of warm-water finfishes-a manual for extension. *FAO Fish. Tech. Pap*. 201.183 p.

Zeitler, M.H., M. Kirchgessner and F.J. Schwarz. 1984. Effects of different protein and energy supplies on carcass composition of carp (*Cyprinus carpio* L.). *Aquaculture* 36: 37-48.

Index

W

Z